INTRODUCTION TO PYTHON AND *SPICE* FOR ELECTRICAL AND COMPUTER ENGINEERS

Butterworth-Heinemann
An imprint of Elsevier

James C. Squire, Ph.D., P.E.
Anthony E. English, Ph.D.

Butterworth-Heinemann is an imprint of Elsevier
125 London Wall, London EC2Y 5AS, United Kingdom
50 Hampshire Street, 5th Floor, Cambridge, MA 02139, United States

---

Notices

Knowledge and best practice in this field are constantly changing. As new research and experience broaden our understanding, changes in research methods or professional practices may become necessary.

Practitioners and researchers must always rely on their own experience and knowledge in evaluating and using any information, methods, compounds, or experiments described herein. In using such information or methods they should be mindful of their own safety and the safety of others, including parties for whom they have a professional responsibility.

To the fullest extent of the law, neither the Publisher nor the authors, contributors, or editors assume any liability for any injury and/or damage to persons or property as a matter of products liability, negligent or otherwise, or from any use or operation of any methods, products, instructions, or ideas contained in the material herein.

ISBN: 978-0-443-19007-0

Acquisition Editor: Stephen Merken
Editorial Project Manager: Aleksandra Packowska
Production Project Manager: Gayathri S
Cover Designer: Matthew Limbert

Printed in the United States of America

Last digit is the print number: 9   8   7   6   5   4   3   2   1

Working together to grow libraries in developing countries

www.elsevier.com • www.bookaid.org

FOREWORD

Welcome, students, as you begin your journey to becoming electrical and computer engineers! You probably chose this major because you wanted a challenging and rewarding college experience. Perhaps you came in search of a field of study that would suit your natural inclinations towards math and science. While some of you may have had prior experiences with First Robotics, Lego League, or a high school electronics course, others may never have considered engineering as a career before a guidance counselor suggested it. Perhaps you have always known engineering was right for you, from the first time you took apart a household gizmo or saved spare parts taken from broken toys. While most people see engineering as a way to help solve society's problems using technology, all of you will soon learn to appreciate our profession as an extended family of colleagues who share an oddball sense of humor, who have actually read their calculator manuals, and who are bothered by technical faults in science fiction movies.

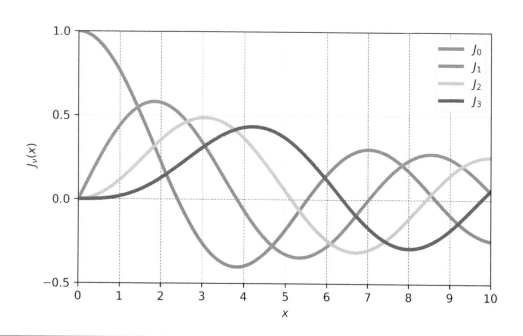

We have designed this textbook not only to help you learn Python programming and Spice simulation skills, but also to introduce you to the culture and profession of electrical engineering. As you progress through the curriculum, you will find that the way you think about problems will change: Your analysis will become clearer, your ability to discard confounding variables will sharpen, and your ability to construct and test models of processes will improve. This will be true whether the problem is how to build an improved voltage amplifier, how to increase the morale of the junior engineers in your unit, or how to grow a company in the face of fierce international competition. Electrical and computer engineering graduates perform well in engineering, leadership, and business roles upon graduation, and many go on to diverse careers in law, medicine, and the military. Indeed, *Forbes Magazine* notes that engineering and computer science majors were found to have higher postgraduate salaries than any other majors.[1] These advantages do not attenuate over time; engineering has long been ranked as the most common undergraduate degrees among Fortune 500 CEOs,[2] more than business and economics degrees.

While a degree in engineering confers many benefits upon graduation, we hope to convey that, in addition to the economic benefits and the opportunity to use technology to help solve societal problems, we also become engineers because we love the challenge of solving difficult problems. That is the thought that keeps both first-year students and seasoned professionals motivated into the early morning hours, and it is a large part of the success and satisfaction you will find in whatever career you ultimately choose.

Welcome to the field.

James C. Squire

Anthony E. English

[1]Somers, Darian, and Josh Moddy. "10 College Majors with the Highest Starting Salaries." *U.S. News & World Report*. 11 September 2019. www.usnews.com/education/best-colleges/sldeshows/10-college-majors-with-the-highest-starting-salaries.
[2]Whitler, Kimbely A. "New Study on CEOs: Is Marketing, Finance, Operations, or Engineering the Best Path to CEO?" *Forbes*. 14 October 2019. www.forbes.com/sites/kimberlywhitler/2019/10/12/new-study-on-ceos-is-marketing-finance-operations-or-engineering-the-best-path-to-the-c-cuite/.

ACKNOWLEDGMENTS

The authors owe a debt of gratitude to their students for their invaluable feedback and the suggestions that they made during the writing of this book. In particular, we wish to thank Shelby Edwards, Elyse Gathy, Conor McCarthy, Timothy Palmer, Alexander Rice, Matthew Steneri, Lily Danforth, and William Weigeshoff. Their reviewers, especially Elisa Barney Smith and David Feinauer, provided excellent feedback from their extensive classroom experience.

James Squire thanks his coauthor, Tony, for managing to keep him on track even under difficult circumstances, for his idiosyncratic sense of engineering humor, but most of all for his never-failing positive attitude; his wife, Laura, whose love and support kept strong even through another year of late-night dinners; and Kevin and Ryan, whose curiosity and enthusiasm for learning he hopes are reflected in this text. He also thanks his parents, James and Marilyn, for putting up with him, his advisors Elazer Edelman and Steve Burns for finding the best in people, and Carl and Gladys Salva for making the best of people.

Anthony English thanks his coauthor, Jim, for his experience, quick wit, and wisdom, which made this project worthwhile despite both of their crazy academic schedules. Anthony would also like to thank his biomedical engineering students for the positive energy they bring into his life. He thanks his family and especially Bonnie for her love, devotion, and understanding over the years, which leaves no words to express the depth of his gratitude.

TABLE OF CONTENTS

Tech Tips

Pro Tips

INTRODUCTION

PURPOSE OF THIS TEXT

We wrote *Introduction to Python and Spice for Electrical Engineers* because we could not find an ideal laboratory manual from among the currently available introductions to Python and introductions to Spice tailored for students of electrical engineering (EE). Many such texts read like user manuals (for instance, "this command does that"), without much attention to connecting with the underlying principles of electrical engineering. Other texts were abstract, focusing more on the heart of procedural programming, but likely too advanced for a first-year engineering audience. A few texts currently available provide excellent introductions to general engineering principles, but they do not focus on the unique concerns of specific majors.

Unlike other manuals, *Introduction to Python and Spice* introduces Python, Spice, and general programming and analysis concepts within an electrical engineering context, and it tailors its content to the specific needs of a single-semester EE course. Thus, it is accessible for first-year students, whether or not they have a background in programming or circuit analysis. While this textbook omits concepts that would be critically important to a computer science student, such as data structures, it includes specialized mathematical applications, such as complex phasor analysis, that are central to frequency-domain analysis in electrical engineering. *Introduction to Python and Spice* also provides an overview of computationally intensive methods used in circuit analysis, electronics, signals and systems, controls, and digital signal processing. It touches on some ideas from the digital courses as well, with the exception of hardware description language (HDL) modeling of general digital systems, which deserves an entire course of its own. In short, this textbook offers both a general introduction to the analog aspects of electrical engineering and a specific overview of the programming packages of Python and Spice.

SOFTWARE FOR ELECTRICAL ENGINEERS

A number of software packages are commonly used in electrical and computer engineering, including Python, Spice, MATLAB, C, C++, VHDL, Java, C#, JavaScript, Multisim, Verilog, Mathcad, and AutoCAD. You will learn about the first two in this text, and many more as you progress through the curriculum.

Python: This relatively new high-level language is meant to be easy to use like MATLAB, provide top-tier support for object-oriented programming like Java and C#, and yet provide access where needed for speed to low-level constructs like C. It is open source, is freely downloadable, and is becoming the standard in certain emerging fields of programming, including machine intelligence.

Spice: Spice is a circuit simulator capable of simulating almost any analog or digital circuit. It lets the user build the circuit graphically and then probe it to read voltages and currents. There are many variants of Spice; in this course, we will use a free version called LTspice, published by Analog Devices.

C: The oldest of the programming languages still widely in use, C's simple syntax is well suited for use in embedded system or miniaturized computers on a chip. It is very closely related to the language used to program Arduino microcontrollers. Despite its ubiquity in miniaturized embedded systems, you will not find it used in full-sized PC programs because it cannot make complex applications as ably as modern programming languages that support object-oriented behavior.

C++: This language, pronounced "Cee-plus-plus," is one of the most common languages used to develop programs that run on personal computers and the Web. Unlike Python and MATLAB, it can run directly from the operating system as a standalone program. Python and MATLAB programs typically must run from within a Python or MATLAB shell. C++ is a superset of the C programming language that adds object-oriented behavior.

VHDL: VHDL is a specialized language used to program a specific type of integrated circuit called a field-programmable gate array or FPGA. These chips do not execute code in a typical fashion. Rather, they run many thousands of processes at the same time, and they require a specialized language to do so. VHDL is slightly more common among defense contractors than its major competitor, Verilog.

MATLAB: MATLAB is commonly used in many fields of engineering, although it is slowly being replaced by Python. MATLAB is a programming language but embeds a very capable graphing package that one of the most popular graphing packages in Python is based upon. MATLAB numerical calculations, especially matrix operations, are its strength. Although it has some symbolic capabilities and extensions, called "toolboxes," it is not intended to be an algebraic solver, like Wolfram Alpha or Mathematica.

Java: Java is a popular multiplatform programming language. Like C++, it is object-oriented and creates stand-alone programs, and many feel it has cleaner syntax than C++. Java is very common, but its overall use is slowly declining.

JavaScript: Despite its name, JavaScript has little in common with Java. This language has grown in popularity since it first became an embedded part of the HTML specification. This is one of the main reasons that the use of Java has declined, and JavaScript is now most commonly found within the HTML that composes Web pages.

C#: Pronounced "C-sharp," this modern, object-oriented programming language is similar to C++, but syntactically far cleaner, like Java. It has become more popular than Java for creating Windows-based programs that run on PCs and the web, but it is tied to the Microsoft operating system.

Multisim: Multisim is another popular circuit simulator, like Spice. It is currently more popular among hobbyists, but it is steadily gaining traction among industry professionals.

Verilog: Like VHDL, Verilog is a language used to program FPGAs. Verilog is more common among nondefense contractors in the United States, and the language looks similar to C.

Mathcad: Mathcad is a hybrid of a spreadsheet and a higher mathematics package, like MATLAB. It is faster to learn, but it also has less programming power. Mathcad is commonly used in Civil Engineering.

AutoCAD, SolidWorks, and *Inventor:* These are solid-modeling programs designed to create virtual models of two- or three-dimensional structures. AutoCAD is very commonly used in civil engineering, while mechanical engineers generally prefer SolidWorks or Inventor. Solid modeling skills are usually not expected among electrical engineering students, but they are useful to have, especially with the proliferation of three dimensional printers.

WHY PYTHON AND SPICE?

While it would be impressive to have student finish their first semester with a working knowledge of all of the above programs, we concentrate on Python and Spice in this book because

1. There is not enough time to cover all the packages and languages used by electrical engineers.
2. Many of the above programs will not make sense until you take more advanced engineering courses. VHDL and Verilog, for example, first require an understanding of state machine and combinational logic, topics taught in digital logic courses.
3. You will likely find Python and Spice helpful in future courses, such as circuits, digital signal processing, and electronics.
4. Concepts learned in Python are transferable to other programming languages you may use in your electrical and computer engineering program.

FORMATTING CONVENTIONS

Each chapter in *Introduction to Python and Spice* has two types of problems: "Practice Problems" and "Lab Problems." Practice Problems will check your general comprehension of specific material from the preceding pages. At the end of each chapter, Lab Problems present more difficult programs that will challenge you to synthesize and implement the material you have learned through the previous chapter.

PRACTICE PROBLEMS

Boxed items like this highlight Practice Problems, which should be completed as you read through the chapter. These relatively simple problems help to significantly increase understanding and retention of the material in a way reading alone does not.

SOLUTIONS

Solutions to all the Practice Problems and the starred (*) Lab Problems are available on the Elsevier website. Python problem solutions are available in a Jupyter Notebook; other problem solutions are in pdf format.

SHADED CALLOUTS

Callouts like this one are useful, but not essential. They include the following:

 Tech Tip: Discussions that link the material to more advanced circuit analysis and laboratory skills

 Pro Tip: Topics relating to the profession of electrical engineering as a whole

 Recall: A review of algebra skills or other materials covered earlier in the text

 Take Note: Particularly important formulas or concepts

 Digging Deeper: Optional, in-depth exploration of topics to enrich your introduction to electrical engineering

PRACTICE PROBLEMS

1. What are the two software packages you will learn about in this course?
2. Name two programming languages that are commonly used in electrical engineering and that you may learn in your career, but that are not taught in this textbook. Describe how they are used.
3. Which of the following advanced electrical engineering courses will build on the computationally intensive methods you will first encounter in this course: circuit analysis, semiconductors, signals and

systems, digital signal processing, microcontrollers, C programming, and/or electronics?

4. Given a choice between C or C# to write a Windows program that would run on a PC, which would you use?

5. What software package would you use to write a program that ran on both Linux and Windows: Java or C#?

6. If you wanted to learn to make prototypes using a rapid-prototyping machine, also known as a three-dimensional printer, would you model them in SolidWorks or AutoCAD, assuming you were planning on sharing the work with mechanical engineering students?

Teacher and Student Resources

Helpful ancillaries have been prepared to aid learning and teaching.

Please visit the student companion site for more details:

https://www.elsevier.com/books-and-journals/book-companion/9780443190070

For qualified professors, additional, instructor-only teaching materials can be requested here:

https://educate.elsevier.com/9780443190070

INTRODUCTION TO PYTHON

1.1 OBJECTIVES

After completing this chapter, you will be able to do the following:

- Download and start Anaconda Navigator
- Use a command line interface to set your working drive and start a Jupyter Notebook
- Set your Jupyter Notebook to your data folder and create a new Python Notebook
- Use Jupyter Notebook command and edit cell modes to enter text and Python code
- Use markdown to self-document a Jupyter Notebook
- Use Python as a basic calculator
- Create variables and then inspect them using the `print` statement
- Perform basic math operations using the `NumPy` package
- Create Python data arrays using the `NumPy` package
- Plot data arrays and shapes using the `Matplotlib` package
- Export your Notebook data and graphics to a word processor or presentation software

1.2 PYTHON TERMINOLOGY

This text teaches programming in Python, an open-source language developed in 1991 by Guido van Rossum that has become one of the world's most popular programming languages. Python is considered an open-source language because it can be freely modified and redistributed by anyone. Python is also considered an interpreted programming language. Python converts the original program statements, called source code, into a lower-level code, called bytecode. The bytecode is then executed by a program called a virtual machine. Using a different virtual machine for each type of computer allows Python to be run on different hardware and operating system platforms. Three common computer operating system platforms are Windows, MacOS, and Linux/Unix.

Python is a specification for a language that can be implemented in different ways. Unlike many other popular languages, there are several different Python language *dialects*. The Python Software Foundation periodically releases the standard distribution of Python, called CPython, at Python.org. Any interpreter that implements the functionality defined by the Python reference outlined on Python.org is considered a Python *implementation*. Alternative implementations of Python, such as Jython (for Java), IronPython (for C#), MicroPython (for microcontrollers), and PyPy, each offer different features and provide greater compatibility with other specific computing languages and hardware.

Python implementations are typically bundled with other supporting programming tools, such as editors and packages, that extend the Python language. Editors allow the user to write Python programs with formatting and code support.

Language extensions, called *packages*, add additional features. We will use two packages, `NumPy` and `Matplotlib`, to add support for complex numeric operations and graphing, respectively. Editors and packages can be bundled into a single convenient installation program; these are called *distributions*. We will use one of the most popular distributions, the *Anaconda distribution*. The Anaconda distribution includes several tools for Python program development.

> **NOTE**
> Python programs are stored in files. A single file holding Python programs is called a *module*. A collection of modules in a directory-like structure is called a *package*.

Anaconda Navigator, thankfully, simplifies the process of installing Python and its various editors and packages. Packages are versioned and may require other packages of specific versions to be installed. Navigator is part of the Anaconda distribution — it not only lets the user specify what editors and packages to install, but also identifies any version dependencies among packages and ensures only compatible updates are installed.

1.3 INSTALLING PYTHON AND ANACONDA NAVIGATOR

Anaconda Navigator can be downloaded from the main Anaconda web page, https://www.anaconda.com/products/distribution. Scroll down to the section that lists the common installation distributions for Windows, MacOS, and Linux, and choose and install the appropriate distribution. Select the default options during the installation.

1.4 STARTING PYTHON FROM A COMMAND SHELL

Run Anaconda Navigator, whose icon is shown in Fig. 1.1. After installing it, PC users will find it on their desktop. Mac users can search for "navigator" using the Mac "spotlight" search tool. Anaconda Navigator will open its standard workspace, shown in Fig. 1.2. The Anaconda Navigator Home window shows many application tools that have been automatically installed or can be added. Spyder, VS Code, and PyCharm are *integrated development environments* (IDEs) that combine Python interpreters, editors, and other tools to support Python software development. The Jupyter Notebook, included in the Anaconda distribution, creates *Notebooks* that combine explanatory text and images with live code.

Python can be run through a *graphical user interface* (GUI) or through a text-based *command line interface* (CLI), commonly called a *command shell*. As end users, we normally use a GUI to navigate through folders, create directories, and delete files. Experienced programmers tend to prefer a command shell since it provides greater control during program development.

ANACONDA
NAVIGATOR

Fig. 1.1 Anaconda Navigator icon.

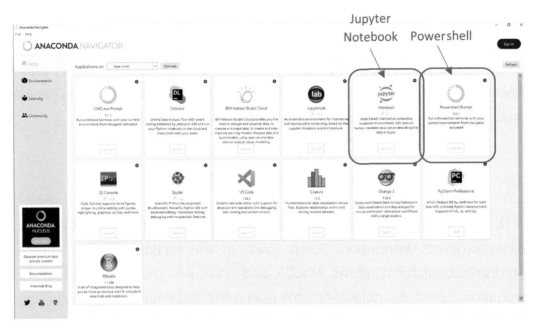

Fig. 1.2 Anaconda Navigator home window. If you are running Windows, after installing and running Anaconda Navigator, a home window will appear similar to the one shown above. If you are running a MacOS or Linux system, the CMD.exe Prompt and PowerShell Prompt applications will be missing.

A Jupyter Notebook is an example of a GUI. Its complexity can be distracting when we are just beginning; therefore we will often introduce new Python commands using the command shell. A common Windows CLI, called a *PowerShell*, can be launched from the Anaconda workspace shown in Fig. 1.2. MacOS and Linux systems use a CLI called a *terminal*.

A command shell allows the user to run Python and then type simple one-line commands to see the result. Need to see what `print("Hello world!")` does? Run it from a command shell. Need to run a 100-line program or write an interactive data analysis program? A Jupyter Notebook is the solution. Command shells are also used as a convenient way to change the current data directory and do other file-oriented operations, such as copying and deleting files.

To launch a command shell on a Windows system, select the PowerShell Prompt application from the Anaconda Navigator Home page shown in Fig. 1.2. If you are running MacOS, launch a Z-Shell Terminal. On a Linux system, use `Ctrl+Alt+T` to launch a terminal. Some commonly used commands are listed below. Try them!

`pwd`	Print the current working directory
`ls`	List the files and folders in the current directory (Linux, MacOS)
`dir`	List the files and folders in the current directory (Windows)

`clear`	Clear the command shell window
`cd subDir`	Change the current directory to the subfolder called `subDir`. After typing `cd`, repeatedly use the Tab key to autocomplete each subdirectory.
`cd ..`	Move up one directory folder level from the current directory
`mkdir newDir`	Create a new directory called `newDir` in the current directory
`rmdir dirName`	Remove or delete the directory called `dirName` in the current directory
`rm fileName`	Remove or delete the file called `fileName` in the current directory
`exit`	Exit the command shell

From the command shell prompt, type `python` to begin an interactive Python session. For a Windows system running a PowerShell, this would appear as

```
(base) PS C:\Users\username>python
```

The Python interpreter should run and return the Python prompt, `>>>`. Test that it is installed correctly using the `print` command. If all is installed correctly, Python will echo "Hello world!" to the display. For example,

```
>>> print("Hello world!")
Hello world!
```

Once you have verified that Python works from the shell, end your Python session by typing `Ctrl+Z` or `exit()` and close the command shell window by typing `exit`. If you are a MacOS user, type `Command+Q` to close the terminal.

1.5 STARTING AND CLOSING A JUPYTER NOTEBOOK

While Python can be run directly from a command shell terminal, it can be used far more flexibly when called from inside a Jupyter Notebook. This can be done by choosing "Jupyter Notebook" from Anaconda Navigator, as highlighted in Fig. 1.2. However, if you wish to save your work to an external drive of your choosing, a simple method is to begin with a shell session, set your working drive there, and then launch Jupyter Notebook directly from the shell. To do this:

1. Launch a command shell as described in Fig. 1.2.
2. Set your working drive if you want to change it from the default, usually the C drive. If, for example, you are working on a Windows system and have a USB in drive E with your

work on it, type `cd e:` followed by a newline. Only set your drive, not the folder, at this step. Note that Windows uses the back slash \ to separate folder names.

```
(base) PS C:\Users\username>cd e:
(base) PS E:\>
```

On MacOS, use the command `ls /Volumes` to list your current drives and `cd /Volumes` to change to that directory. From there use the `cd` command to enter your external drive. If your drive name has spaces in it, surround the name by double quotes. On a Linux system, `lsblk` will list your block devices, including USB drives. External drives are typically found in the `/media` directory or one of its subdirectories. Note that MacOS and Linux use the forward slash / to separate folder names.

3. Launch a Jupyter Notebook directly from the command shell by typing `jupyter notebook` at the prompt.

```
(base) PS E:>jupyter notebook
```

This will bring up the Jupyter Notebook dashboard in a browser, as shown in Fig. 1.3. Do not close the command shell window during your session!

Next, create a data folder to store your data and programs, and make it the current working directory:

1. Select the folder hyperlinks on the left side of the screen to navigate to where you want to create your data folder.
2. Click on the New button highlighted in Fig. 1.3 on the right side of the screen, and then choose "Folder" from the drop-down menu to create your data folder.
3. Rename the folder `Chapter 1 Python` by selecting it (click the box to the left of its name) and then click the Rename button that will appear above the selection boxes.
4. Enter the folder. Click the New button, and then choose "Python" from the drop-down menu to create your first Jupyter Notebook running Python code.
5. Test that the Notebook communicates with Python by typing `print("Hello World!")` from the prompt. Send it to the Python interpreter by pressing `Shift+Enter` or hit the run button: ▶ Run

Your Jupyter Notebook should appear as shown in Fig. 1.4.

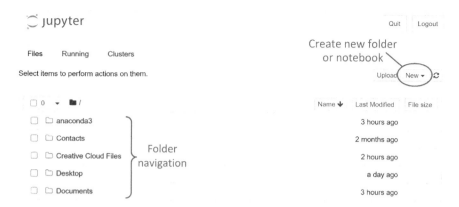

Fig. 1.3 Representative Jupyter Notebook dashboard. Your dashboard's appearance will depend on the folders you have in your selected drive. From the Jupyter Notebook dashboard, you can set the working folder directory and start a new Notebook or open an existing Notebook.

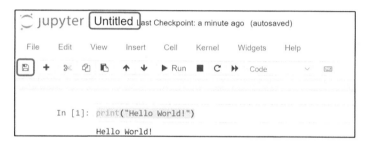

Fig. 1.4 Python "Hello World" program. Change "Untitled" to the name of the file you want to save it as. Click on the highlighted save icon to store your file.

Save your Notebook and exit Jupyter Notebook and Anaconda Navigator.

1. Click the Untitled link at the top of Jupyter Notebook highlighted in Fig. 1.4. Name it `Chapter 1 Notebook`.
2. Save your work by clicking the save icon highlighted in Fig. 1.4.
3. Under the File menu, choose "Close and Halt." Then close the Jupyter window tab, the shell window, and Anaconda Navigator.

1.6 JUPYTER NOTEBOOK ANNOTATED CODE CREATION

Reopen the `Chapter 1 Notebook` saved earlier.

1. Run Anaconda Navigator.
2. Use a command shell to set your drive location if you are using an external drive for your work.
3. Launch a Jupyter Notebook.
4. Navigate to your Chapter 1 Notebook and open it.

1.6.1 Code and Markdown Cells

Your work is contained in cells. There are two types of cells, *code cells* and *markdown cells*.

- Code cells contain code to be executed. Executing, or running, a code cell produces its output below the cell.
- Markdown cells contain text with formatting instructions that can document your code. When a markdown cell is run, the plain text it contains is rendered as stylized text. For example, the * symbol will be rendered as a bulleted • symbol.

Both code and markdown cells can be in *command* or *edit* mode. Command mode remaps the keyboard to issue commands that affect the entire Notebook or blocks of cells. For example, typing c in this mode will copy the current cell and v will paste it. Edit mode allows one to type Python code inside a cell.

- Command mode: Left-click to the left of the cell, or press Esc. Cell border turns blue.
- Edit mode: Left-click inside the cell. Cell border turns green.

Try toggling back and forth between command and edit modes in a cell by hitting Enter and Esc. Notice how the active cell color turns blue and green as shown in Fig. 1.5.

Ctrl+Shift+P opens a list of Jupyter Notebook commands. A short summary of some useful operations when in command mode is as follows:

- Press H to open a popup window displaying all the hotkeys in this list and more!
- Insert a cell above or below the current cell by pressing A or B, respectively.
- Transform the cell to markdown by typing M.
- Transform the cell to Code by typing Y.
- Typing D twice deletes the current cell.
- Undo cell deletion by pressing Z.
- Scroll up or down the cells by using the arrow keys, ↑ and ↓, respectively.

Fig. 1.5 Code cells in command mode, *left* (blue), and edit mode, *right* (green). Notice that the cursor appears in edit mode ready for your typed input.

- To select multiple cells, hold the `Shift` key down and select the cells or use the up and down keys. Use `Shift+M` to merge multiple selected cells. To split a cell at the cursor, use `Ctrl+Shift+-` in edit mode.

1.6.2 Self-Documenting Jupyter Notebooks Using Markdown

Jupyter Notebooks can hold more than code; they can also include *markdown*, which is text designed to read by a human, not by the Python interpreter. It allows the programmer to document what a section of code does, or to notate a homework assignment, as shown in Fig 1.6.

Homework Set 1

Problem 1: Using Python as a Calculator

```
[1]: result = 1/(2**(1/2))
     print("My favorite number is %0.4f" %(result))

     My favorite number is 0.7071
```

Fig. 1.6 Markdown text. This example shows a mixture of markdown and code text. The bolded "Homework Set 1" and "Problem 1: Using Python as a Calculator" are differently formatted markdown texts that explain what the code beneath it does.

To create markdown text that describes an existing code cell, create a new cell above the code and enter markdown mode. Do this by selecting your code cell, entering command mode by pressing `Esc`, typing `A` to create a new cell above it, and typing `Esc`, `M` to turn it into a markdown cell. Give your code a level 1 heading by typing

```
# Hello World Example
```

The markdown `#` sign is a directive to make the following text on the current line a level 1 heading. Run the cell using `Ctrl+Enter` or by pressing the run button. You can re-edit your markdown text by placing the cursor in the cell and double-clicking. In markdown notice that there is no `In [ ]:` next to the cell.

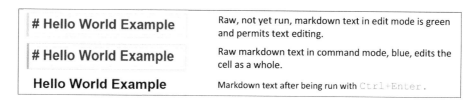

# Hello World Example	Raw, not yet run, markdown text in edit mode is green and permits text editing.
# Hello World Example	Raw markdown text in command mode, blue, edits the cell as a whole.
Hello World Example	Markdown text after being run with `Ctrl+Enter`.

Fig. 1.7 Raw and processed markdown text. The *top* shows markdown text while in edit mode, the *middle* while in command mode, and the *bottom* after the cell is run.

Markdown cells have many useful formatting directives, such as the following:

- `#` creates a level 1 heading
- `##` creates a level 2 heading
- `**text**` creates **bolded** text
- `*text*` creates *italicized* text
- `* item` creates a • bullet for lists
- Paragraphs are separated by single blank lines
- Form lists using a `*` symbol on separate lines or form ordered lists with numbers on each line.

1.7 USING PYTHON AS A CALCULATOR

Practice using the self-documenting markdown commands from the last section as we explore how to use Python as a calculator. Open a new cell below your current work by placing the cursor in the last cell, entering command mode by pressing the `Esc` key, and then typing `B` to open a new cell below it. While in command mode, turn it into a markdown cell by typing `M` (alternatively, select "Markdown" from the toolbar beneath the menu) and enter the following text to create a new level 1 heading:

```
# Using Python as a Calculator
```

Press `Ctrl+Enter` or the run button to format the text. Make sure the current cell is a code cell, enter `2/5`, and run it. The output should be of the form shown in Fig. 1.8.

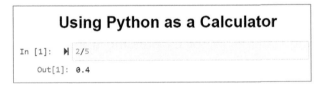

Fig. 1.8 Output produced from a markdown cell and code cell.

Repeat this for the statements

```
3 + (4*9) ** 20
x = 5
x + 32
x = 2
y = x + 1
```

To observe the values of *x* and *y*, use the print statements `print(x)` and `print(y)`. You can also just type `y` in a separate cell to observe its value.

Python can be used as a sophisticated calculator. It recognizes the basic commands

```
+

-

*

/

**
```

As in many programming languages, addition, subtraction, multiplication, and division are represented by `+`, `-`, `*`, and `/`. Python uses `**` to represent exponents, so `10**6` means 10^6. Python uses the standard order of operator precedence to evaluate expressions, so `1+3**2` is the same as $1+ (3^2) = 10$, not $(1+3)^2 = 16$. When in doubt, use parentheses to be clear.

Python is like a high-end scientific calculator in several ways:

1. Python only operates on the current line. When the value of `x` was changed from 5 to 2 in the previous example, unlike a spreadsheet, but like a calculator, it did not change the previous calculation involving `x`, but only changed current and future calculations.
2. When there are variables involved, such as `x` and `y` in this example, and an equals sign, Python first evaluates the statement to the right of the equals sign, and then it sets the variable on the left to that number. Most programming languages also follow this convention.

Python is unlike a calculator in other ways, which may be confusing at first. For example,

1. Python cannot perform symbolic algebra without importing special packages. Unlike many modern calculators, which allow one to enter `x**2-1 = 0` and solve for x, Python does not understand how to work with variables that are not set to a numerical value. Thus Python will understand `x = 4+7` because it will set x to a specific number. It *assigns* the variable on the left to the expression on the right. It will understand `y = x + 5` and set y to 16. However, it will not know what to do with `2*y = 3*x`. If you want to solve for y in this expression, you must do it yourself using algebra, and then enter `y = 3*x/2`.

2. Python does not understand implicit multiplication. The expression `4(3+1)` will generate an error message, unlike `4*(3+1)`, which will evaluate to 16.

PRACTICE PROBLEMS

For every practice problem followed by the © symbol, record only the Python command that solves the problem. For problems followed by the ® symbol, write the result that Python returns.

    1.1 Using Python, calculate 2^{10}. ©®

    1.2 Using Python, calculate $2(6/13) + \frac{2.17 - 3.29}{2}$. ©®

TECH TIP: CHARGE AND CURRENT

Two important quantities in Electrical and Computer Engineering (ECE) are current and charge. An accurate way to think of these quantities is the analog of water being pumped through pipes around a closed circuit as shown in Fig. 1.9.

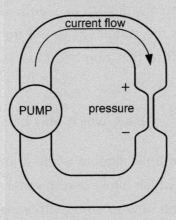

Fig. 1.9 Hydraulic model of charge and current. Electric current flow is like water flow in a pipe. The amount, or volume, of water is like the amount of charge.

Charge

The amount of water, measured in gallons, is like electrical charge, measured in coulombs or "C". Engineers typically use the variable "Q" to stand for charge, such as $Q_1 = 0.023$ C.

Current

The quantity of water flowing past a point, measured in gallons per second, is like electrical current, measured in amps or "A". Engineers use the variable "I" to stand for current, such as $I_2 = 12$ A.

Charge and Constant Current

We can intuitively understand the relationship between a steady current flow and the amount of charge that passes. For example, if 2 gallons per second of current passed a point in 3 seconds, there would be 2 gal/s · 3 s = 6 gal of water that had flowed. Similarly, if a constant current $I = 2$ A flowed in a wire for 3 seconds, the charge passed would be $Q = 2$ A · 3 s = 6 C. For a constant current flow of I amps over a period of t seconds, the general equation for total current Q in coulombs is $Q = I \cdot t$, shown graphically in Fig. 1.10.

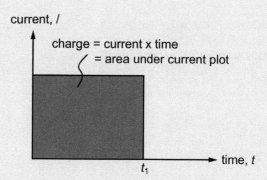

current, I

charge = current x time
= area under current plot

time, t

t_1

Fig. 1.10 Current versus time plot. The area under the curve represents the charge.

Notice that instead of the formula above, the charge that flows from time 0 to t_1 is also equal to the area under the current line between $t = 0$ and $t = t_1$. This is true even if the current changes with time. Think about this using the water analogy until it becomes obvious.

DIGGING DEEPER

Those who have taken calculus will recognize that the formula for constant current, $Q = i \cdot t$, generalizes to

$$Q = \int_0^{t_1} i(t)\, dt$$

for changing current $i(t)$. The current Q is still equal to the area under the current graph.

PRACTICE PROBLEMS

1.3 Using Python, find the total charge that flows through a circuit with a 2.5 A current source that has been turned on for 45 seconds. (Remember units!) ©®

1.4 A car's headlights were left on overnight for 8 hours. If the headlights draw 0.74 A, use Python to calculate how much charge they drew overnight. ©®

NOTE

Periodically save your Notebook using `Ctrl+S`.

PRO TIP: IEEE

Electrical engineers and other technologists can join the Institute of Electrical and Electronic Engineers (IEEE), an international professional society pronounced "I triple-E." This society has thousands of sections around the world that meet monthly to discuss topics of current engineering interest. These meetings are popular among working engineers to network, socialize, and stay current in the field. Most universities sponsor an IEEE student

> branch, which may arrange interuniversity robotics competitions, lead field trips, and help students find internships and, ultimately, jobs. Find out more at *ieee.org*.

1.8 VARIABLES

Begin a new Python session — this time, for variety, from the command shell prompt (click the command shell prompt, then type `python` from the prompt). Python can make it easier to perform complex calculations by using variables. For example, if a design equation says the value of resistor R_1 can be found from the equation

$$R_1 = \frac{1}{2\pi f C} - R_2 \tag{1.1}$$

and we know $f = 2000$, $C = 20 \cdot 10^{-9}$, and $R_2 = 3000$, one can type

```
>>> R1 = 1/(2*3.14159*2000*20e-9) - 3000
>>> R1
978.8769381109569
```

However, it is clearer and less error-prone to type

```
>>> f = 2000
>>> C = 20e-9
>>> R2 = 3000
>>> pi = 3.14159
>>> R1 = 1/(2*pi*f*C)-R2
>>> R1
978.8769381109569
```

This last example illustrates several important points.

1. Why we use variables. It makes the problem statement more natural. It also makes it easy to rerun the problem with different values. To see how $f = 100$ changes the

calculation, for example, just change the *f* line to `f = 100` and then re-evaluate `R1 = 1/(2*pi*f*C) - R2`.

2. The number pi. One cannot enter Greek characters directly, but one can define a variable and set it equal to the value of π. Later, we will see that there are also built-in constants that can be used.

3. Python does not understand units without special packages, so use `R2 = 3000`, for instance, and not `R2 = 3000 Ω` or `R2 = 3 kΩ`.

4. Python does not use subscripts or superscripts, so a variable such as R_1 must be written `R1`.

1.9 NAMING AND INSPECTING VARIABLES

Variable names must begin with a letter or an underscore. They may be any length, and capitalization matters. Electrical engineers tend to name resistors as R, R_1, R_2, R_3, capacitors as C, C_1, C_2, frequency as f, and time as t. Note that Python considers R1 and r1 to be two different variables. Variables may be composed of lowercase and uppercase letters, numbers (but not as the leading character), and the underscore _ character.

Examples of valid variable names include

```
R27
f
index
Whats_the_airspeed_velocity_of_an_unladen_swallow
```

Invalid variable names include

```
3R
spent$
done?
```

Once a variable is created, it stays defined unless you restart the kernel, shut down Python, or explicitly delete the variable. The kernel is what executes code contained in a Notebook. Every time you start a Python session, whether from the command shell or from a Jupyter Notebook, it starts "clean," with no user-defined variables from previous sessions. During a Jupyter Notebook session, you can restart the kernel and begin again with a clean slate. Notice that restarting the kernel resets the `In [ ]` variables. The `In [ ]` labels note the order in which each statement was executed. Fig. 1.11 illustrates restarting the kernel in a Jupyter Notebook.

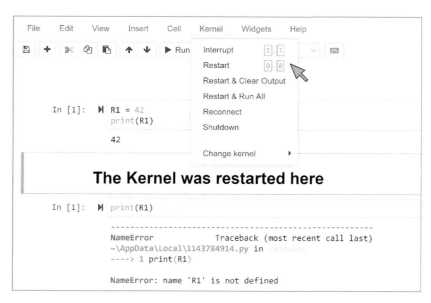

Fig. 1.11 Restarting the kernel clears all previously defined variables.

It is rare to need to delete a variable, but in case you do, it is done using the `del` command. For example,

```
>>> del R27
```

deletes the variable R27. To find what variables are currently defined in a Jupyter Notebook, evaluate `%who`. Typing `%whos` will list the defined variables with their types and values. Fig. 1.12 illustrates the application of both the `%who` and `%whos` commands. Commands that begin with

```
In [1]:   1  f = 2000
          2  C = 20e-9
          3  R2 = 3000
          4  pi = 3.141592
          5  R1 = 1/(2*pi*f*C) - R2

In [2]:   1  %who

          C     R1     R2     f     pi

In [3]:   1  %whos

          Variable   Type    Data/Info
          ------------------------------
          C          float   2e-08
          R1         float   978.874405078699
          R2         int     3000
          f          int     2000
          pi         float   3.141592
```

Fig. 1.12 Using the `%who` and `%whos` magic commands to list the defined variables.

a `%` sign in a Jupyter Notebook are called *magics* and are not part of the Python program language but are very useful. To get a complete list of magics in a Jupyter Notebook, type `%` `lsmagic.`

PRACTICE PROBLEMS

Which variables are valid? If they are not valid, explain why.

1.5 TheAnswerIs42

1.6 SetPtEmBeR21

1.7 2B_or_Not2B

1.8 R1

1.9 Current&7

1.10 PARENTHESIS AND IMPLIED MULTIPLICATION

Unlike most calculators, Python does not understand implied multiplication. Be careful to avoid entering commands like the ones shown in Fig. 1.13. These commands will return an error message. Instead, multiplication must be explicitly written out with a `*`, as shown in Fig. 1.14.

```
>>> 4 (3 + 7)

or

>>> a = 7
>>> 3 a
```
Do not do this!

```
In [1]:  4(3+7)
-------------------------------------------------
TypeError                              Traceback
(most recent call last)
~/ipykernel_20100/1589794347.py in <module>
----> 1 4(3+7)
TypeError: 'int' object is not callable
```

Fig. 1.13 Using implied multiplication with Python statements produces errors.

```
>>> 4 * (3 +7)
40
>>> a = 7
>>> 3 * a
21
```
Fixed

```
In [1]:  4*(3+7)
Out[1]: 40

In [2]:  a=7
         3*a
Out[2]: 21
```

Fig. 1.14 Explicitly writing out multiplication with Python statements fixes the errors.

Use Python to solve the following problems. ®

1.10 $4\pi\sqrt{\frac{2}{7}}$ Hint: Use ** (1/2) for the square root function.

1.11 $\dfrac{1}{\dfrac{1}{R_1} + \dfrac{1}{R_2}}$ If $R_1 = 6$ and $R_2 = 12$.

TECH TIP: VOLTAGE, CURRENT, CHARGE, AND RESISTANCE

We already presented an analogy between water flowing in a closed loop of pipes and electric current flowing through a closed set of wires. The amount of water in gallons flowing past a point over a given time interval is like the charge in coulombs that flows past a point in a wire in a time interval. The amount of water flow, measured in gallons per minute, is like the amount of charge flow, called current, which is measured in coulombs per second, or, more conveniently, in amps. To complete the analogy, the pump that creates a pressure difference in the water to make the current flow is like a voltage source. The pressure difference between any two points is like voltage difference, and the resistance to the flow of water in the piping system is like a resistor. Table 1.1 provides a summary of electrical circuit properties and their water analogs.

TABLE 1.1   Summary of electrical and water circuit analogs

Quantity	Abbreviation	Units	Water analogy
Charge	Q	Coulombs (C)	Water volume, measured in gallons
Current	I	Amps (A)	Water flow, measured in gallons per second
Voltage	V	Volts (V)	Pressure difference
Resistance	R	Ohms (Ω)	Narrow pipe causing resistance to water flow

To represent a circuit, we need the components listed in Table 1.2. A simple electrical circuit and its water analogy are shown in Fig. 1.15. Try to visualize the electrical current flow though the circuit, as it is pressurized to flow by the voltage source and loses that pressure as it squeezes through the resistor.

TABLE 1.2 Electric circuit components

Component	Schematic symbol	Water analogy
Wire	————————	Pipe
Voltage source: e.g., battery		Pump creating a pressure difference
Resistor		Narrow pipe resisting water flow

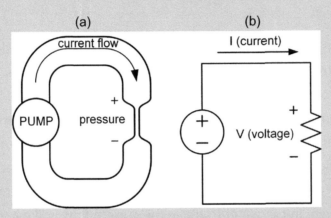

Fig. 1.15 Pumped water system and analogous electric circuit. (a) The pump creates a pressure difference to encourage water to flow through a restricted pipe. (b) This is analogous to a voltage source such as a battery that creates a voltage difference, causing electric current to flow through a resistor. The resistances in the pipe and circuit are both unavoidable and necessary to prevent an infinite current from flowing.

PRACTICE PROBLEMS

1.12   Using the analogy of the pumped water system, would current increase or decrease as pipe resistance increased, assuming the pump pressure remained constant?

About two centuries ago, Georg Ohm summarized the relationship between current, voltage, and resistance with the formula named after him. The formula is called Ohm's Law and is

$$V = IR \qquad (1.2)$$

where V is the voltage, I the current, and R the resistance. If we try to understand this relationship in terms of the water analogy, it conveys two different ideas.

1. Voltage is proportional to current. As more water current is forced through a narrowed pipe, a greater pressure will develop across it.
2. Voltage is proportional to resistance. As the pipe narrows and its resistance increases, water flowing through it will also cause a greater pressure to develop across it.

There are two other formulations of Ohm's Law, one solved for current,

$$I = \frac{V}{R} \qquad (1.3)$$

and one solved for resistance,

$$R = \frac{V}{I} \qquad (1.4)$$

PRACTICE PROBLEMS

1.13 A 68 Ω resistor has 2.4 A of current pumped through it. Use Python to compute the voltage across the resistor. ©®

1.14 That same 68 Ω resistor is now connected to a 120 V voltage source. Use Python to compute the amount of current that flows. Use Python to compute the current flow. ©®

1.11 IMPORTING MODULES AND PACKAGES

Using Python's interactive interpreter from the command shell is the easiest way to start. Simply type each command, one line at a time, after the prompt $>>>$. The Python interpreter will execute each line. For example,

```
>>> print("Hello World")
Hello World
>>> 2 + 3
5
```

Although the interpreter will evaluate each line as entered, Jupyter Notebooks will allow you to group many commands into a single cell and evaluate them all at once, as summarized in Fig. 1.16.

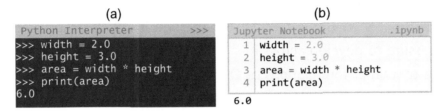

Fig. 1.16 Entering Python commands at the interpreter and in a Jupyter Notebook code cell. (a) When a PowerShell is used, the Python interpreter executes each command as entered. (b) A Jupyter Notebook code cell, however, allows one to group multiple commands into a code cell and execute them all at once. Python commands can also be grouped into a file, typically having a .py ending, called a module.

Grouped commands can be saved into files called *modules*. Modules play an important role in Python program development. Each module can perform specific tasks, and saving well-tested modules helps the programmer reuse and share them.

Several modules can be grouped together into a directory-like structure called a *package*. A package can also contain subpackages as a directory can contain subdirectories. There are many premade modules and packages to do a variety of tasks, such as mathematics, plotting, and data reading. For a module or package to be used, it must be installed and then imported. Fortunately, the Anaconda distribution installs many commonly used packages automatically. One of these is NumPy, which stands for Numerical Python. As the name suggests, it extends Python's ability to perform mathematics.

The NumPy package contains mathematical constants and many basic mathematical functions such as square roots, exponentials, and powers. It is customary to use the NumPy package by typing

```
>>> import numpy as np
```

The term `np` is called an *alias,* and it makes accessing the variables and functions in the package easier. To access the square root function and the constant π in the `NumPy` package, for example, one would use the command `np.sqrt(2)` and `np.pi`, respectively.

The differences among a line of Python code run interactively, a module, and a package are shown below.

The *Python Interpreter* executes one line at a time. For example, type the following from the command line.

```
>>> print("Hi")
Hi
>>> 2 + 3
5
```

A series of Python commands saved in a file ending in .py is called a *module.* When the file is run, all the commands are executed.

```
width = 4.2
height = 5.0
area = width * height
print(area)
```

More complex programs import *packages* such as `NumPy`. A module is not run alone; it is imported into scripts or other modules to provide new commands or variables.

```
import numpy as np
radius = np.sqrt(2)
area = np.pi * radius **2
print(area)
```

1.12 SPECIAL SYMBOLS AND MATH FUNCTIONS

Electrical engineers use imaginary numbers very frequently, for instance, $3 + 7j$, where $j = \sqrt{-1}$. This is different from the way most mathematicians write it, $3 + 7i$. Engineers use j because i is reserved to mean current. Some engineering texts place the symbol j in front of the number rather than behind it, but Python uses the j convention with the number in front of the j, as demonstrated in Fig. 1.17.

```
>>> 3 + 7j
(3+7j)
```
```
In [1]:  1 3 + 7j
Out[1]: (3+7j)
```

Fig. 1.17 Complex numbers using the Python interpreter *(left)* and a Jupyter Notebook code cell *(right).*

Another constant that electrical engineers use surprisingly often is the ratio of a circle's perimeter to its diameter, or π. In Python, the number π can be accessed from the `NumPy` package. To use the number π, it is common to import the `NumPy` package with the alias `np` and access π as shown in Fig. 1.18.

```
>>> import numpy as np
>>> np.pi
3.141592653589793
```

```
In [1]:    1  import numpy as np
           2  np.pi

Out[1]:  3.141592653589793
```

Fig. 1.18 The number π using both the Python interpreter *(left)* and a Jupyter Notebook code cell *(right).*

1.13 FORMATTING NUMBERS

Internally, Python keeps track of all numbers with about 16 digits of precision and displays all digits by default. To show output using fewer digits, a formatted print statement can be used. An example of a formatted statement for a fixed-point number with three digits of precision is shown in Fig. 1.19.

$$\underbrace{\#\#.\underbrace{\#\#\#}}_{6.3f}$$

Fig. 1.19 Formatting fixed-point numbers in Python. The first number is the minimum total number of digits, including the decimal point. It will first print spaces, if necessary, to print this width. The second number represents the number of digits after the decimal point. The letter `f` tells the `print` statement that this is a fixed-point number.

Using the interpreter, one could use formatting in a `print` statement: for example,

```
>>> x = 21.12345
>>> print(f"{x:6.3f}")
21.123
>>> y = 123.123
>>> print(f"{y:0.2f}")
123.12
```

A formatted string starts with an `f`, and the string is surrounded by quotation marks, `"`. The formatted variable is enclosed by curly brackets, `{}`, and the variable and format are separated by a colon, `:`. Formatting strings will be covered in more detail in Chapter 4.

In electrical engineering, we tend to report three significant digits in our answers, even if we can compute more numerically. This is because the components we use — resistors, capacitors, and inductors — rarely have more than three significant digits of accuracy in their values, and the instruments we use to measure voltage and current in undergraduate laboratories are usually limited to about three significant digits as well.

DIGGING DEEPER
Curious as to why Python keeps *about* 16 digits of internal precision, rather than exactly 16 digits? Python displays numbers in base 10, but internally works with them in base 2. It stores them in exactly 52 base-2 digits. In base 10 this is $\log_{10}(2^{52})$ or 15.6536 digits.

1.14 SCIENTIFIC NOTATION

Very large and very small numbers are more easily read in scientific notation. For example, a typical capacitor used in our field is a $6.8 \cdot 10^{-12}$ farad capacitor. It would be much harder to pick one out from the parts cabinet if its label read 0.0000000000068. In Python, one can enter this value in scientific notation using

```
>>> 6.8e-12
```

Notice that there is no space between the `6.8` and the `e`. This is not implied multiplication, which Python does not do, but rather a shortcut that gives the same result as entering `6.8*10**(-12)`. Yet, it is easier to enter and read. Internally, this is also faster to process for Python. Later in this book we will introduce engineering notation, which is an even more compact way to write these values than scientific notation.

Formatted printing in scientific notation follows the same pattern as fixed-point printing except that the `f` is replaced with an `e`. For the above example,

```
>>> x = 0.0000000000068
>>> print(f"{x:9.3e}")
6.800e-12
```

1.15 EXPONENTS AND THEIR INVERSES: exp, **, sqrt, log, log10

Electrical engineers frequently use the natural exponential function e^x. This function can be found in the `NumPy` package along with square root, natural log, and log base-10. For example, to compute e^{-1} in Python, rather than typing `2.718**(-1)`, one would type

```
>>> import numpy as np
>>> np.exp(-1)
0.36787944117144233
```

The first line is needed to import the `NumPy` package. In this case, the `NumPy` package has been imported with the alias `np`. Using the alias `np`, with a dot extension, allows access to the `NumPy` package variables and functions. The inverse of the natural exponent is the natural logarithm. Some mathematics textbooks call this ln, as in ln(2), but in the Python `NumPy` package it is called log, as in `np.log(2)`. For example,

```
>>> import numpy as np
>>> np.log(np.exp(-3.5))
-3.5
```

Electrical engineers often use base 10 logarithms when analyzing signals whose amplitudes vary widely. The amplitude of a signal, for example, is commonly measured in decibels, which involve base 10 logarithms. To find a logarithm with the base 10, such as $\log_{10}(10,000) = 4$, use the Python command `np.log10()`, as in

```
>>> import numpy as np
>>> np.log10(10000)
4.0
```

Raising a number, other than e, to various powers is done using the `**` symbol. For example, to represent 2^{32} in Python type

```
>>> 2**32
4294967296
```

The square root of a number can be found in Python with the `np.sqrt()` function. For example, $\sqrt{65536}$, another number commonly used in EE, is found by typing

```
>>> import numpy as np
>>> np.sqrt(65536)
256.0
```

> **RECALL**
>
> Need to find a root other than a square root? Cube roots and higher are the same as taking the reciprocal exponential. For example,
>
> $\sqrt[7]{1024} = 1024^{\left(\frac{1}{7}\right)}$ in Python is written as `1024**(1/7)`. Python's `**` function will let you find any root.

PRACTICE PROBLEMS

1.15 Use Python to evaluate e^{-5}. ©®

1.16 Use Python to evaluate $20 \log_{10}\left(\frac{1}{\sqrt{2}}\right)$. ©®
This number, surprisingly close to an integer, turns up frequently in filter design problems. Hint: Using the `NumPy` package with the alias `np`, the Python command for a square root is `np.sqrt(x)`. ©®

1.16 TRIG FUNCTIONS AND THEIR INVERSES

Electrical engineers frequently use trigonometric functions, especially cosines, to represent signals. Python's trig functions are provided by the `NumPy` package. After importing the `NumPy` package with the alias `np`, one can use `np.cos()`, `np.sin()`, and `np.tan()` to find the cosine, sine, and tangent of a number given in radians. For example, $\cos(\pi)$ is found by typing

```
>>> import numpy as np
>>> np.cos(np.pi)
-1.0
```

Inverse trig functions are given by the `NumPy` functions `arccos()`, `arcsin()`, and `arctan()`, where the result is returned in radians. For example, the inverse cosine in radians of −0.5 is found using

```
>>> import numpy as np
>>> np.arccos(-0.5)
2.0943951023931957
```

RECALL
There are 360 degrees and 2π radians in a circle. The conversion factor must therefore be the ratio, $180/\pi$ or $\pi/180$. There are more degrees in an angle than radians, so to convert to degrees multiply by $180/\pi$. There are fewer radians than degrees in an angle, so to convert to radians multiply by $\pi/180$.

Although all the `NumPy` trigonometric functions use radians, engineers often use degrees. There is a Python conversion function called `radians()` that accepts degrees and returns radians. To use it to compute cos(60°), for example, type

```
>>> import numpy as np
>>> np.cos(np.radians(60))
0.5000000000000001
```

DIGGING DEEPER
The cosine of 60° is 0.5, not 0.0.5000000000000001. Python internally uses floating point operations in base 2 for numerical calculations. Small errors may occur either from the iterative numerical algorithms themselves or in converting between base 2 and base 10 to display results. This can cause much larger errors when two almost-identical quantities are subtracted, called catastrophic cancellation. As an example, this operation should be 0 but is not:

```
(np.cos(np.radians(60))-0.5).
```

Similarly, one can convert from radians to degrees using the `NumPy degrees()` function. For example, to find arcsin(1/2) with the answer in degrees, type

```
>>> import numpy as np
>>> np.degrees(np.arcsin(1/2))
30.000000000000004
```

1.17 CREATING ARRAYS

So far, Python variables have been scalars, that is, single numeric values such as a = 57. Python variables can also be arrays that consist of collections of numbers stored, for example, as a one-dimensional array or a two-dimensional array with rows and columns. We have already used the NumPy trigonometric functions and predefined constants such as π. The NumPy package also contains the functions needed to work with arrays. For example, to represent a vector $y = [12, 3, -45, 2.7, \pi]$ as a one-dimensional NumPy array and display it, type

```
>>> import numpy as np
>>> y = np.array([12, 3, -45, 2.7, np.pi])
>>> print(y)
[ 12. 3. -45. 2.7 3.14159265 ]
```

Arrays with regularly spaced numbers can be created easily using the NumPy linspace() function. We will explore this and other ways to create arrays more fully in the next chapter, and in Section 1.18 we will use it to plot functions. One way to call it is with the syntax np.linspace(start, stop, num_values). For instance, to create the array z that begins at 0, ends at 1, and has 6 values, that is, [0 0.2 0.4 0.6 0.8 1], use

```
>>> import numpy as np
>>> z = np.linspace(0,1,6)
```

NOTE

A common mistake is to think that one can create a list of 10 integers from 0 to 10 with the command np.linspace(0,10,10). There are actually 11 numbers in the array 0, 1, 2, 3, 4, 5, 6, 7, 8, 9, 10, so it is done with the command np.linspace(0,10,11). This is called the "fence post error" after the old trick question, "how many fence posts are needed to cover 10 feet if they are placed a foot apart?" You know the answer is 11, not 10.

Most common mathematical functions will automatically operate on each of the numbers within the array, such as exp(), log(), cos(), and the others we have discussed so far. Using the five-element y array defined earlier, we can compute the cosine of each of the five members of y and add 1 to it in one step using

```
>>> import numpy as np
>>> y = np.array([12, 3, -45, np.pi, 2.7])
>>> x = np.cos(y) + 1
>>> print(x)
[1.84385396 0.0100075 1.52532199 0. 0.09592786]
```

DIGGING DEEPER

You will notice that the fourth number in the *x* array above printed as "0." rather than simply "0" (without decimal points). This is because there are some real numbers in the array, so Python makes all of them real. Had the array been defined using only integers, for instance, `y = np.array([0, 1, 2])` then `print(x)` would display them as `[0, 1, 2]` without a decimal point.

PRACTICE PROBLEMS

1.17   Using `NumPy` operations, create the array *x* = [1 4 5 8], add 5 to it and then multiply the result by 3, call the result *y*, and print it to the screen. ©®

Random numbers can also be created quickly using the `np.random.rand()` function. A vector array with *N* random numbers, each of which is between 0 and 1, can be created using `np.random.rand(N)`. For example,

```
>>> import numpy as np
>>> N = 3
>>> x = np.random.rand(N)
>>> print(x)
[0.14124364 0.94290792 0.37152498]
```

Your random numbers will, in all probability, be different.

TECH TIP: USING A DIGITAL MULTIMETER (DMM) TO MEASURE STEADY VOLTAGE

Modes: Modern DMMs, such as the one shown in Fig. 1.20, may seem imposing at first sight, but you will be using them like a pro. To measure a voltage, first set the selector switch to DC voltage. Possible modes usually include

- AC (varying) voltage, marked with $\sim$
- DC (constant) voltage, marked with $\mathrel{\overline{\cdots}}$
- AC current
- DC current
- Resistance

Range: DMMs may autorange, like the one in Fig. 1.20, or require manual adjustment. If autoranging, a DMM will automatically move the decimal point in the measurement to make the reading as precise as possible. Manual-ranging DMMs split each type of measurement (voltage, current, etc.) into different ranges. For a manual-ranging meter, select the lowest voltage that includes the one you expect to measure. To measure a 1.5 V battery, for example, if the DMM offers maximum voltage ranges of 0.2, 2, 20, and 200 V, choose the 2 V range.

Fig. 1.20 Autoranging DMM.

Probe Jacks: Plug the probes into the correct jacks. DMMs have one red probe and one black probe, and there are usually three or more jacks to place them. The black probe goes into the jack labeled "COM" or "Common," and the red probe goes into the jack labeled "V" or "Volts."

Probe Tips: The red probe goes to the more positive end, and the black tip to the more negative end. For the DMM pictured here, the battery, nominally marked 1.5 V, reads 1.78 V. If the probe tips were reversed, so that the red probe were at the negative end of the battery, the DMM would read −1.78 V.

1.18 PLOTTING DATA

Python has a comprehensive set of plotting abilities, many of which we will explore in Chapter 3. Python `Matplotlib` is a plotting library that, like `NumPy`, is installed with the Anaconda distribution. Most of the `Matplotlib` commands are in the `pyplot` submodule and are usually imported under the `plt` alias using the statement `import matplotlib.pyplot as plt`. To display plots within a Jupyter Notebook, use the command `%matplotlib inline`. Fig. 1.21 demonstrates a simple cosine plotting program. Remember to use `Shift+Enter` to evaluate the Jupyter Notebook cell.

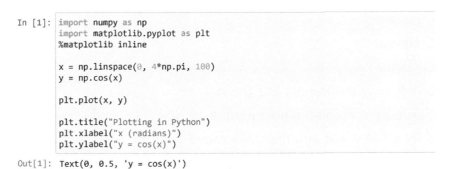

```
In [1]: import numpy as np
        import matplotlib.pyplot as plt
        %matplotlib inline

        x = np.linspace(0, 4*np.pi, 100)
        y = np.cos(x)

        plt.plot(x, y)

        plt.title("Plotting in Python")
        plt.xlabel("x (radians)")
        plt.ylabel("y = cos(x)")
```

Out[1]: Text(0, 0.5, 'y = cos(x)')

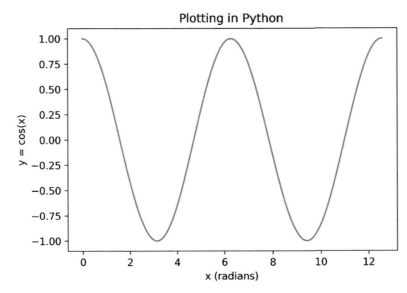

Fig. 1.21 Plotting with Python. The `Matplotlib` package allows data to be plotted and labeled.

To save the plot image, right-click the image and choose "save image." If you want to copy and paste it into a Microsoft Word document, right-click the image, choose "copy image," and then paste it into the Word document with `Ctrl+V`. To save your Python code in a Word document, select the code by clicking and dragging with the mouse, `Ctrl+C` to copy it to the clipboard, and then `Ctrl+V` to paste. The entire Notebook can also be saved in different formats by choosing "File" and then "Download As" from the drop-down menu to select a file format.

PRACTICE PROBLEMS

1.18 Describe in words what each line of code does that creates Fig. 1.21.

1.19 Plot a triangle with vertices at $x = [0\ 1\ 0.5]$ and $y = [0\ 0\ 0.7]$. Hints: Recall that the Matplotlib library's plot command plots two arrays given as horizontal points first and then vertical points. The example in Fig. 1.21 used `linspace` to generate the points, but you can also define them explicitly, as explained in Section 1.17. You may think you need just three points to define your triangle, but you will need a fourth to close it, where the final point is a copy of the starting point. To copy and paste the figure into a Microsoft Word document, place your cursor in the plot image area, right-click and choose copy image, and then paste the image into your word document by pressing `Ctrl+V`.

1.19 GETTING HELP

This text is a study guide, not a Python reference manual. A reference manual is an exhaustive listing of what every command does, but it does not provide instruction on how to use the system as a whole. In contrast, a study guide teaches the reader by introducing commands in small, sequential units that build upon each other. The `help()` method calls the built-in Python reference manual. To get help on the `print()` function, for instance, execute `help(print)` as shown in Fig. 1.22. You can get more comprehensive help by using the help button on the main menu and selecting a topic.

1.20 SAVING AND LOADING ARRAY DATA

Python does not remember variables between sessions. That is, when Python starts, it has no variables defined. Once `NumPy array` variables are defined, you may want to save them so that

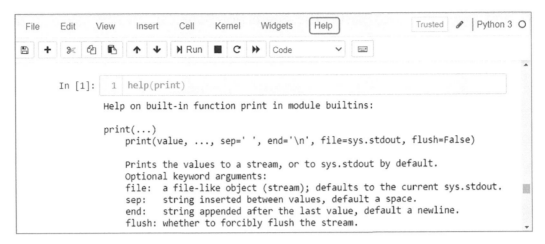

```
In [1]:  1  help(print)

         Help on built-in function print in module builtins:

         print(...)
             print(value, ..., sep=' ', end='\n', file=sys.stdout, flush=False)

             Prints the values to a stream, or to sys.stdout by default.
             Optional keyword arguments:
             file:  a file-like object (stream); defaults to the current sys.stdout.
             sep:   string inserted between values, default a space.
             end:   string appended after the last value, default a newline.
             flush: whether to forcibly flush the stream.
```

Fig. 1.22 Using the `help()` function. The `help() function` displays information on the input argument.

they can be loaded during another session. Variables worth saving are typically large arrays holding many data values. For example, an array defined to hold all the close-of-day-stock prices for IBM would have over 20,000 numbers in it.

The `NumPy` package has a function called `savetxt()` that saves a `NumPy array` in a txt file. A txt file uses a format, called ASCII, that you can read and edit in a program such as Notepad. For example, the program

```
>>> import numpy as np
>>> t = np.linspace(0.0, 1.0, 11)
>>> v = np.sin(2.0 * np.pi * 1.0 * t)
>>> np.savetxt('myData.txt', v)
>>> np.savetxt('myDataInRows.txt', [t, v])
>>> np.savetxt('myDataInColumns.txt', np.transpose([t, v]))
```

saves generated sinusoidal data in three different txt files. The first data file, `'myData.txt'`, consists of a single column of the `array v` data. The second file, `'myDataInRows.txt'`, has the data stored as two rows, the first row for `array t` and the second row for `array v`. The last statement uses the `NumPy transpose()` function to interchange the rows and columns so the data file, `'myDataInColumns.txt'`, has the array data stored in columns instead of rows. Note that these files are saved in your current working directory.

If you close and restart Python, thus clearing all variables from memory, you can reload your data into NumPy arrays using the NumPy loadtxt() function. For example, to load a single data column from the txt file myData.txt into the NumPy array v, use

```
>>> import numpy as np
>>> v = np.loadtxt ('myData.txt')
```

Similarly, to load two rows of data from the txt file 'myDataInRows.txt', the first into the NumPy array t and the second into the NumPy array v, type

```
>>> import numpy as np
>>> t, v = np.loadtxt('myDataInRows.txt')
```

If you have your data stored as two columns in the file 'myDataInColumns.txt', use the NumPy transpose() function to switch the rows and columns; that is, execute

```
>>> import numpy as np
>>> t, v = np.transpose(np.loadtxt('myDataInColumns.txt'))
```

To save and read complex data, use the dtype = complex attribute in the NumPy loadtxt() function. For example,

```
import numpy as np
zdata = np.array([4+4j, 1+1j]) # create array zdata
np.savetxt('z.txt',zdata) # save into z.txt
zdata2 = np.loadtxt('z.txt',dtype = complex) # read into zdata2
```

This is also a simple way to import data generated by other applications into Python NumPy arrays.

> **NOTE**
>
> Python uses the hashtag or pound sign, #, to define a single-line comment. Everything following the # symbol on the same line is ignored by the interpreter. Comments help the reader understand your code and will be discussed in more depth in Chapter 4.

1.21 PYTHON COMMAND WINDOW KEYBOARD SHORTCUTS

When using a command shell, not a Jupyter Notebook, very often you need to repeat a sequence of keystrokes exactly or almost exactly. This has probably happened to you already, and you will find these keyboard shortcuts very convenient. To recall previous commands, press the up-arrow key ↑. Keep pressing to go further back in the command history. Once the command is visible, you can edit it or run it again by pressing `Enter`.

To select previous cells in a Jupyter Notebook, enter command mode, by pressing `Esc` if in edit mode, and then press the up-arrow key ↑ to select the cells that you wish to change. Either change them directly in those cells and rerun them with `Shift+Enter`, or copy, `Ctrl+C`, and paste, `Ctrl+V`, them into new cells that you can insert before or after the current cells with the hotkey `A` or `B`, respectively.

1.22 EXPORTING YOUR NOTEBOOK RESULTS

Once you have completed your analysis, you may want to save your Notebook results in a format other than a Jupyter Notebook. The entire Notebook can be downloaded in different formats from the File menu, or you can use copy and paste to transfer specific regions of text to a word processor or presentation package, as described in Section 1.18. This will not work for graphic items such as plots made with the `Matplotlib` package. For now, the simplest method of copying `Matplotlib` graphics, although low-resolution, is to take a screenshot of the plot and insert it into the word processor or presentation application. In Windows, this is `Windows+PrtScn`, and in MacOS, it is `Command+Shift+4`. Linux screenshot tools depend on the Linux flavor. Ubuntu, for example, is bundled with GNOME Screenshot. In Chapter 3 we will introduce another method for rendering `Matplotlib` graphics as scalable vector graphics files, .svg, or high-resolution portable network graphics files, .png.

PRO TIP: CAREERS IN ELECTRICAL ENGINEERING

Careers in electrical engineering can be examined by sector, by title, or by subdiscipline. "Sector" refers to who hires the engineer, such as a company, the government, or a university. "Title" refers to the job description, which is a function of the sector. "Subdiscipline" refers to the specific electrical engineering field, such as microelectronics, power, or telecommunications. There are three primary sectors that hire electrical engineers, private industry, government, and academia (in order of the number of annual hires), and each has unique benefits. Salaries tend to be highest in private industry, while government positions generally offer the greatest job security, and engineers working in academic institutions enjoy significant independence in choosing research topics.

Each sector offers different types of job titles. Private industry has a great variety, including

- Research and Development: Develop protypes of new devices and technologies;
- Production: Optimize manufacturing processes to reduce cost;
- Sales: Work as part of a team to sell a technological solution to a client;
- Project Management: Manage teams of engineers and other specialists as they solve problems too broad in scope for a single engineer; and
- Consulting: Help other companies solve technological problems, often involving production and project management.

Government jobs often involve

- Project Management: Oversee private industries performing government contract work; and
- Testing: Determine if products meet safety and health requirements, which can be difficult to ascertain when the technology under review is novel.

Academic positions typically include

- Research and Development: Often with a longer time horizon to implementation than R&D positions in private industry; and
- Teaching: Both undergraduate and graduate students.

Any employment sector can be involved with any of the main subdisciplines of Electrical Engineering. Major EE subdisciplines include

- Microelectronics: Fabrication of extremely small integrated circuits;
- Power: Generation and transmission of large amounts of electrical power over long distances;
- Controls: Use programmable logic or microcontrollers to precisely control motion or heat;
- Signal Processing: Design systems that measure voltage, flow, pressure, or temperature;
- Telecommunications: Build systems that transmit data by wire, optical fiber, or radio waves;
- Computer Engineering: Design of computers and associated hardware.

PYTHON COMMAND REVIEW

Basic Math Operations

`+ - * /` No implied multiplication, so 2*y* is entered `2*y`.

`**` Power. For example, 2^8 is entered `2**8`.

Exponent Operations

First requires `import numpy as np`.

`np.sqrt()` Square root.

`np.log()` Natural log.

`np.log10()` Log base 10.

`np.exp()` Natural exponent.

Predefined Constants

First requires `import numpy as np`.

`1j` $\sqrt{-1}$, for instance, as `3+7j`. Use `j`, not `i`, and put number first.

`np.pi` π.

Formatting Numbers

`68e-12` Type of scientific notation, here representing 68×10^{-12}.

`print(f'{x:6.4f}')` Formatted print of *x* using fixed point number with four decimal places.

`print(f'{x:9.3e}')` Formatted print of *x* using scientific notation with three decimal places.

Trigonometric Functions

First requires `import numpy as np`.

`np.cos()` Cosine function given argument in radians.

`np.sin()` Sine function given argument in radians.

`np.tan()` Tangent function given argument in radians.

`np.arccos()` Inverse cosine function returning argument in radians.

`np.arcsin()` Inverse sine function returning argument in radians.

`np.arctan()` Inverse tangent function returning argument in radians.

`np.radians ()` Given an argument in degrees returns value in radians.

`np.degrees ()` Given an argument in radians returns value in degrees.

`np.cos(np.radians(6))` Find cos(6°), by first be changing degrees to radians.

`np.degrees(np.arctan(1))` Find arctan(1) in degrees, since the trig functions use radians.

Saving and Loading Data

First requires `import numpy as np`.

`np.savetxt('data.txt',x)` Save variable *x* in file called data.txt.

`x=np.loadtxt('data.txt')` Load real data from data.txt into variable *x*.

`z=np.loadtxt('data.txt',dtype=complex)` Load complex data from data.txt into variable *z*.

Array Functions

	First requires `import numpy as np`.
`np.array([2, 4, -6])`	Creates a vector with components 2, 4, and −6.
`np.linspace(0, 5, 3)`	Creates three linearly spaced numbers from 0 to 5, that is, [0 2.5 5].
`np.random.rand(1,5)`	Creates a 1 row × 5 column array of random numbers from 0 to 1.

Plots

	First requires `import matplotlib.pyplot as plt`.
`%matplotlib inline`	Required to display plots in a Jupyter Notebook.
`plt.plot(x, y)`	Plot x and y vector arrays.
`plt.title()`	Add a title to the plot.
`plt.xlabel()`	Add a label to the x axis.
`plt.ylabel()`	Add a label to the y axis.
`#`	Python code comment. Everything after `#` is ignored by the interpreter.

COMMAND SHELL REVIEW

`pwd`	Print the current working directory.
`ls`	List the files and folders in the current directory (Linux, MacOS).
`dir`	List the files and folders in the current directory (Windows).
`clear`	Clear the command shell window.
`cd subDir`	Change the current directory to the subfolder called subDir.
`cd ..`	Move up one directory folder level from the current directory.
`mkdir newDir`	Create a new directory called newDir in the current directory.
`rmdir dirName`	Remove or delete the directory called dirName in the current directory.
`rm fileName`	Remove or delete the file called fileName in the current directory.
`exit`	Exit the command shell.

PYTHON INTERPRETER COMMAND REVIEW

Start Python interpreter	1. Launch a command shell or terminal.
	2. Type `python` to start the interpreter, changing the prompt to $>>>$.
Close Python interpreter	`Ctrl+Z` or `exit()`.
↑	Recalls the last command.
↑↑	Recalls the next to last command, etc.

JUPYTER NOTEBOOK COMMAND REVIEW

Start Jupyter Notebook	1. Run the PowerShell prompt `C:\Users\username>`.
	2. Change to the desired data drive, not folder, such as `D:`
	3. Type `jupyter notebook`.
Run a cell's contents	`Shift+Enter` or press the Run button.
Close Jupyter Notebook	1. File menu, "Close and Halt."
	2. Close the Jupyter window tab.

Code and Markdown Cells in Command and Edit Modes

Fig. 1.23 Code and markdown cells

`Esc`	Puts active cell in command mode (or click in left margin).
`Enter`	Puts active cell in edit mode (or click in cell).

Command Mode

`A`	Inserts new cell above the current cell and makes it the active cell.
`B`	Inserts new cell below the current cell and makes it the active cell.
`Y`	Makes the active cell a code cell.
`M`	Makes the active cell a markdown cell.
`Shift+M`	Merges all selected cells.
`H`	Displays summary of common hotkeys.

Markdown Mode

`#`	Creates a level 1 heading.
`##`	Creates a level 2 heading.
`**bold**`	Makes text between the pair of `**` **bold**.
`*italic*`	Makes text between the pair of `*` *italic*.
`*`	Creates a bulleted list using a `*` on a separate line.
`1., 2.,`	Creates a numbered list using a number on a separate line.

LAB PROBLEMS

These problems require more thought than the embedded practice problems that were designed to check general comprehension. For every practice problem followed by a © symbol, record only the Python *command* that solves the problem. For problems followed by a ® symbol, write the *result* that Python returns. An example of a question and answer with both a © and an ® requirement is

Question: Use Python to evaluate $4 + \sqrt[6]{5}$. ©®

Answer: Command: $4 + (5**(1/6))$

                    Result: 5.30766048601183

Solutions to all starred (*) problems are available on the Elsevier website (refer to page xix for the link).

1.1 Computers use sets of binary digits, or bits, to represent information. A bit can be 0 or 1, so a single bit can store two unique states. Two bits can store four states, namely 00, 01, 10, or 11. Many microprocessors store information in sets of 8 bits, called bytes.

   a) How many unique states can be represented in a byte? Hint: It is much more than 16. ©®

   b) Most modern desktop computers store information in sets of 64 bits, called 64-bit words. How many unique states can that represent? ©®

1.2* A common circuit analyzed in future courses is the voltage divider, which will be discussed more thoroughly in the Tech Tip following Section 2.13. It is associated with the equation

$$V_1 = V_{in} \frac{R_1}{R_1 + R_2} \qquad (1.5)$$

   a) Use Python to define the variables V_{in}, R_1, and R_2, equal to 12 V, 68 Ω, and 12 Ω, respectively. Calculate V_1. Hint: Do not use dimensions such as volts and Ω in Python, but remember to write them in your result. ©®

   b) Repeat the above calculation but with $R_1 = 13$ Ω. Notice how defining variables can make recalculation easier. ©®

1.3 Modern music synthesizer design leans heavily on signal processing theory. The note called A4 has a frequency of 440 Hz (cycles per second). Each "semitone" above it is $\sqrt[12]{2}$ times higher in frequency than the note before it. For example, one semitone higher than A4 has a frequency of $440 \sqrt[12]{2} = 466.16$ Hz, and two semitones higher has a frequency of $440 \sqrt[12]{2} \sqrt[12]{2} = 440 \left(\sqrt[12]{2} \right)^2 = 493.88$ Hz. Use Python to calculate the frequency of the note 7 semitones higher than A4. Musicians would call that note E5. Format the result neatly using the `print()` command. ©®

1.4* Use Python to calculate $\sqrt{3} + \sqrt[3]{4}$. Hint: Read the shaded "Recall" box at the end of Section 1.15. ©®

1.5 Use Python to determine what famous irrational number, used frequently in signal processing, the following continued fraction approximates:

$$3 + \cfrac{1}{7 + \cfrac{1}{16}} \tag{1.6}$$

1.6* One way to approximate the irrational number e, frequently encountered in control theory, signal processing, and many other areas, is by using the equation

$$\lim_{n \to \infty} \left(1 + \frac{1}{n} \right)^n \tag{1.7}$$

a) Using Python's array functions, evaluate this limit for $n = 100$, 1000, and 10,000 using the variable n. The first command should be `n =` `np.array([100, 1000, 10000])`. Complete the problem using one additional command using arrays. ©®

b) Find the % error = (approximate − exact)/exact x 100% for $n = 100$, 1000, and 10,000. Remember that the exponential function is `exp()`. Do this in a single command using arrays. Notice how using variables simplifies the work. ©®

1.7 Using Python, find the total charge that flows through a circuit with a 2.5 A current source that has been turned on for 45 seconds. Remember to give units in your answer! ©®

1.8 Standard alkaline AA batteries can deliver 0.02 A of current to power an LED light for 120 hours (surprising, but true!). How much charge circulates through the battery in this time? Remember that the current/time relationship in the Tech Tip: Charge and Current at the end of Section 1.7 assumes time is measured in seconds. ©®

1.9 An electric car motor draws 720 A from a 350 V battery when under full load. Use Python to determine

a) What is the equivalent resistance of the motor? ©®

b) How much charge flows through the motor in 2 minutes? Be careful of the time units. ©®

1.10* Current flowing through a circuit linearly drops over time from a maximum of 2.7 A, when it is first turned on, to 0 A, ten seconds later. How much charge has flowed through the circuit during those 10 seconds? Draw a diagram of current versus time first, and then use Python to calculate the answer. ©®

1.11 If you cut a wire and separate the ends in the schematic shown in Fig. 1.24, all current stops flowing. What is the equivalent of cutting the wire in the pipe analogy? Hint: It is not just cutting the pipe because then water would flow everywhere.

Fig. 1.24 Circuit and water analogy (Lab Problem 1.11).

1.12 A 120 V voltage source is connected across a resistor and 6.8 A of current flows. Use Python to compute the value of the resistor. ©®

1.13* A 12 Ω resistor is connected to a 14 V source. Use Python to compute how much charge flows through the resistor after 2 seconds. Hint: You will need two formulas. ©®

1.14 A student learning Python attempts to solve an Ohm's Law problem for the voltage caused when $I = 2.3 \times 10^{-6}$ A flows through an $R = 45$ Ω resistor by typing the three lines of code

```
I = 2.3EE-6
R = 4500 ohms
V = IR
```

Unfortunately, there is one mistake in every one of the student's three lines. Rewrite each of the three lines correctly and then use Python to calculate the answer. ©®

1.15 A ⅛ W slow-blow fuse is rated to blow (open) if ¼ A flows through it for 15 seconds, or ½ A flows though it for 3 seconds, or ¾ A flows through it for 1/8 second. Here, W is a unit of power, just as A is the unit of current. Write the Python array expression that calculates the total charge that flows through the fuse before it blows in each of these three scenarios. Use array variables for both the currents and the times. Perform the calculation using a single array calculation, not three separate ones. ©®

1.16* Engineers use some unusual numbers; e and π are both irrational numbers with values about 2.72 and 3.14, respectively, and the imaginary number $j = \sqrt{-1}$ is so unusual that even its discoverer, Euler, said it was merely a mathematical curiosity of no practical use. But these three numbers combine in a surprising way that later courses will put to good use. Use Python to evaluate $e^{j2\pi}$ and write the result as simply as possible. Extremely small deviations from integers come from machine inaccuracies and can be rounded safely. Hint: The answer is surprisingly neither complex nor irrational. ©®

1.17 Imaginary numbers are commonly used in electrical engineering. Find the result of the expression

$$\left(\frac{6j\sqrt{7}}{4+5j}\right)\left(\frac{\sqrt{28}}{4-5j}\right) \tag{1.8}$$

using Python. Extremely small deviations from integers come from machine inaccuracies and can be rounded safely. Hint: The exact answer, after extremely small values are rounded, is purely imaginary. ©®

1.18 The amount of power that can be dissipated by a typical carbon film resistor of length L and diameter D, shown in Fig. 1.25, is given by the equation

$$P = k\left(\pi D^2 + 2\pi(L - D)\right), \tag{1.9}$$

where L and D are in mm, and P is in W.

a) Use Python to find k if $D = 2.5$ mm and $L = 6.5$ mm for a ¼ W resistor. ©®

b) Use a single Python `NumPy` array to compute the power ratings of three resistor sizes, all with $D = 5$ mm and with $L = 13$, 17, and 25 mm, using the value of k calculated above. ©®

Fig. 1.25 Carbon film resistor dimensions (Lab Problem 1.18).

1.19* A student learning Python tries to evaluate $\frac{16^{1/2}}{2}$ by typing

```
16 ** 1 / 2 / 2
```

However, this command returns 4.0, an incorrect answer. What should the student have typed, and what is the correct answer? ©®

1.20 Use Python to find the cosine of $\pi/6$ radians. ©®

1.21 Use Python to evaluate the cosine of $60°$. ©®

1.22* Use Python to calculate $23 \cos(38.5°)$. ©®

1.23 Johnson noise is a type of electrical noise generated in all resistors by internal thermal fluctuations. Although relatively small, it can be noticeable when very small voltages are measured, especially over high-value resistors in hot environments. It is calculated using the equation

$$\overline{V_n^2} = 4\,k_B \cdot T \cdot R \qquad (1.10)$$

where

$\overline{V_n^2}$ is the power of the Johnson noise per Hz bandwidth measured in units of V^2/Hz

$k_B = 1.38 \cdot 10^{-23}\,J/K$

T = temperature in K

R = resistor value in Ω

Use Python to find $\overline{V_n^2}$ for a 620 Ω resistor at temperatures ranging from 275 K to 300 K in 5 K increments. Do this using arrays so it can be done with a single command. The answers should be very, very small, less than billionths of a V^2/Hz. ©®

Additional challenge: Do the above using Python's built-in `range()` function. Place it inside a `NumPy` array like this: `np.array(range(...))`. Use `help(range)` to learn more.

1.24 Create Python variables R1 = 22, R2 = 47, and R3 = 68 and then evaluate the expression

$$\frac{1}{\dfrac{1}{R_1} + \dfrac{1}{R_2} + \dfrac{1}{R_3}} \qquad (1.11)$$

which calculates the equivalent resistance of all three resistors in parallel. ©®

1.25* Evaluate the complex expression

$$\left| \frac{j26}{20 - j26} \right| \qquad (1.12)$$

The vertical lines enclosing the complex expression indicate that the magnitude of the complex expression is taken. This will result in a real number. To do this, you will need to read the help for the `abs()` command. ©®

1.26 A student attempts to solve the expression $R_1\sqrt{5}$ for $R_1 = 36\ \Omega$ with the Python code

```
import numpy as np
R1=36
1answer = R1*np.Sqrt[5]
```

There are three errors in the code. Write down the corrected code and then use Python to compute the correct answer. ©®

1.27 Use Ohm's Law, Python, and your knowledge of algebra to find the resistance between two dry fingers of one hand if, when 22×10^{-6} A of current is forced to flow in one fingertip and out the other, well below the ability of a person to sense it, 0.62 V is measured between the fingertips. Format your answer using Python's `print()` command with a variable such as, for example, R = 123.45 Ω. ©®

1.28* After correctly importing `NumPy` with the `np` alias, a student attempts to calculate $4 \cos(36°)$ using the expression

```
ans1=4 np.cos[36]
```

There are three errors in this expression. Write the correct expression and the correct answer. Hint: The answer should be between 3 and 4. If your answer is negative, reread Section 1.16: Trig Functions and Their Inverses. ©®

1.29 The *sinc* function often occurs in fields of signal processing, among others, and is defined as

$$y = \frac{\sin(x)}{x} \tag{1.13}$$

Use Python and the `np.linspace()` function to plot 200 points of Eq. 1.13 as x varies between -20 and 20. Hint: To plot, remember first use the commands

```
import matplotlib.pyplot as plt
%matplotlib inline
```

©®

1.30* Create a plot, with the x axis extending from 1 to 50, of 50 random numbers whose values are plotted on the y axis. It should look like a jagged line extending horizontally. Each of the 50 numbers should be between 0 and 1.

Hints: See the hints for Problem 1.29. Use `np.linspace()` to create an x vector from 1 to 50, and `np.random.rand()` to create your y vector with the 50 random numbers. Show the commands you used and include the plot. ©®

1.31 Plot the first letter of your last name. It may be very rough. If your last name begins with "S," for example, a jagged reversed "Z" shape is fine as long as it is recognizable. Hints: See the hints for Problem 1.29. ®

PYTHON AS A CALCULATOR

2.1 OBJECTIVES

After completing this chapter, you will be able to use Python as an advanced calculator to perform the following operations:

- Create built-in Python data objects and determine a data object type
- Round a number and separate it into integer and fractional components
- Build and manipulate vector arrays
- Perform vector array mathematics
- Use complex numbers to analyze signals and systems
- Transform complex numbers from rectangular to polar formats to analyze circuits
- Create and manipulate strings
- Build and manipulate matrices
- Solve simultaneous equations to analyze circuits

2.2 BUILT-IN PYTHON DATA TYPES

Built-in Python data objects can be classified as scalar or nonscalar. The scalar data types, summarized in Table 2.1, represent single data type entities, while nonscalar types represent collections of scalar types. The scalar Python objects are int, float, complex, bool, and None. The three Python types int, float, and complex represent numeric data. The type int is used for integers such as -4 or 563. The float type represents floating point numbers, which have a decimal point, such as 3., 2.718, or 1.38e-23. The complex type represents complex numbers such as 2j or 1.4+6.1j. The type bool represents Boolean values that can be either True or False. The keyword None is used to define a null value or no value at all. None is not 0 or False. The data type None is simply a NoneType.

TABLE 2.1   Built-in Python scalar data object types

Type	Description	Examples
Numeric Types		
int float complex	Integer value Floating point number Complex number	2, 5, 1005 2., 42.4, 1.2e4 1.4+2.1j
Boolean Type		
bool	Boolean	True, False
NoneType		
None	NoneType. Not 0 or False... just None	None

Nonscalar Python data objects, summarized in Table 2.2, include sequence and nonsequence data types. The sequence data types are list, tuple, and str. A list type is a sequential list of scalar objects surrounded by square brackets, [], that can be modified. A tuple is a sequential list of data objects surrounded by round brackets, (), that cannot be modified. The data type str represents a sequence of characters. A set is a nonordered collection of objects that can be modified or changed, surrounded by curly brackets, {}. A dictionary data type consists of data pairs where the first element, called a key, is associated with another element, called a value, separated by a colon {key:value}. Additional data types can be created or imported using packages. The NumPy package, for example, allows collections of the data array type, called the ndarray, to be created.

2.5.3 Accessing and Changing Values in Vectors

Python handles vectors like many programming languages in two important ways:

- It indexes vectors starting from 0, not 1. The first value in vector v is at index 0.
- Python uses square brackets [ ], not parentheses ( ), to index vectors.

Thus, the Python command to read the fourth value in vector v and set variable x equal to it is

```
>>> x = v[3]
```

To set the second value in vector v to π, type

```
>>> v[1] = np.pi
```

You can retrieve multiple values at once by providing a set of indices. This process, called array slicing, uses the general format `[start:stop:step]` to specify a set of equally spaced indices separated by intervals of size `step` that begin at `start` and up to, but not including `stop`. If the `step` size is left out, it defaults to one. If `start` is missing, it defaults to zero, and if `stop` is missing, it defaults to the number of vector elements.

For example, to retrieve the third (index two) through seventh (index six) values of `v = np.array([11, 12, 13, 14, 15, 16, 17, 18, 19, 20, 21])`, type `v[2:7]`. This returns the vector `np.array([13, 14, 15, 16, 17])`. Do you see why? Note that the returned vector includes the `start` index but not the `stop` index. The `[2:7]` defined a set of indices used to retrieve those values of v. Study the following examples and understand how the set of indices are defined.

```
>>> import numpy as np
>>> v = np.arange(11,22)
>>> print(v)
[11 12 13 14 15 16 17 18 19 20 21]
>>> print(v[2:7])
[13 14 15 16 17]
>>> print(v[2:])
[13 14 15 16 17 18 19 20 21]
>>> print(v[:4])
[11 12 13 14]
>>> print(v[1:8:2])
[12 14 16 18]
```

2.5.4 Indexing Relative to the End

To specify the first *n* elements of a vector `v`, use `v[0:n]` or just `v[:n]`. For example, to access the first five elements of the vector `v`, use `v[0:5]` or `v[:5]`.

But how can you specify all elements from the fourth from last index through to the second from last index? You can use the negative sign to refer to the index from the end. Fig. 2.2 shows both the positive and negative indices associated with a vector array `v`. Note that there are a total of 11 elements in the array; that is, `len(v)` is 11.

For example, to generate the vector `v`, shown in Fig. 2.2, and print the vector subarray `[18 19 20]`, type

```
>>> import numpy as np
>>> v = np.arange(11,22)
>>> print(v)
[11 12 13 14 15 16 17 18 19 20 21]
>>> print(v[-4:-1])
[18 19 20]
```

−11	−10	−9	−8	−7	−6	−5	−4	−3	−2	−1
11	12	13	14	15	16	17	18	19	20	21
0	1	2	3	4	5	6	7	8	9	10

Fig. 2.2 Positive and negative indices. The positive indices begin at 0, starting with the first element, `11`. The negative indices start at −1, beginning with the last element, `21`. The data values, shown in the boxes, can be generated using `v = np.arange(11,22)`.

2.5.5 Removing Vector Array Elements

What does "removing an element" in a vector mean? We already know how to set it to zero. For example, setting the first element in the vector `v = np.array([1, 2, 3])` to `0` can be done with the command

```
>>> v[0] = 0
```

To remove the value, however, means to shorten the vector length. This is accomplished using the `NumPy delete()` command. In the example above, typing

```
>>> w = np.delete(v, 1)
```

returns a vector `w = np.array([1, 3])` equal to `np.array([1, 2, 3])` but without the second element. The length of `w` is 2. You can also specify a set of indices of elements to remove. For example,

```
>>> np.delete(v, [0, 2])
```

deletes the first and third elements. If you want to change the vector `v` itself, use the command

```
>>> v = np.delete(v, 1)
```

2.5.6 Building Vectors from Other Vectors

Vectors can be assembled from other vectors and scalars. If `x = np.array([1, 2, 3])` and `y = np.array([6, 7, 8])`, then typing

```
>>> z = np.concatenate([x, [7.5, 8.5], y, [3.1]])
```

will create a vector `z` with the components `[1. 2. 3. 7.5 8.5 6. 7. 8. 3.1]`. A common mistake is to omit the square brackets just inside the round brackets. Also note that even a single number, such as 3.1 in this case, must be enclosed in square brackets.

2.5.7 Inserting Elements in Vectors

To insert 5 at the start of the vector `v = np.array([1, 2, 3])`, type the command

```
>>> v = np.insert(v, 0, 5)
```

To insert a `7` at the end of vector `v = np.array([1, 2, 3])`, type the command

```
v = np.insert(v, len(v), 7)
```

Vector `v` now holds `np.array([1, 2, 3, 7])`.

To insert a `5` into the middle of vector `v = np.array([1, 2, 3, 4])`, at index 2, type the command

```
v = np.insert(v, 2, 5)
```

The vector `v = np.array([1, 2, 5, 3, 4])` now has a 5 in the middle at index 2.

You could also use the concatenate function to perform the same operation using indexing:

```
v = np.concatenate([v[0:2], [5], v[2:]])
```

Notice the extra set of square brackets just inside the round brackets. A common mistake is to omit them.

PRACTICE PROBLEMS

2.4 Given `x = np.arange(20,31,2)`, write the Python commands to
1. Find out how many numbers vector `x` holds. ©
2. Display the contents of `x`. ©
3. Retrieve the second value of the vector `x`. ©
4. Set vector `y` to equal the second through fifth elements of `x`. ©
5. Set vector `w = x` but with a 1 inserted into the front and end of vector. That is, `w = np.array([1, 20, 22, 24, 26, 28, 30, 1])`. Create `w` using the `x` vector. ©

PRO TIP: BUYING A DIGITAL MULTIMETER (DMM)

The digital multimeter is an important tool for practicing electrical engineers and one you will use frequently. Unlike relatively expensive oscilloscopes, a good DMM costs relatively little and will last a long time. If you decide to purchase your own DMM, look for the following features (in addition to the resistance and AC/DC voltage and current measuring that all DMMs should have):
- Sound continuity (buzzes when the probes are connected so you can check for shorts and opens while looking at your work—very helpful for debugging!)
- Diode checking
- Capacitance measuring
- Frequency measuring (optional)

Other features are more expensive, and you are less likely to need them as you start out. These features include

- True RMS measurement
- NIST-traceable calibration
- Go/no-go testing
- Bar graph displays

Name brand instruments such as Fluke tend to be very rugged and accurate, but they also cost more. One brand used by the authors, shown in Fig. 2.3, is about 5 years old and costs less than a high-end student engineering calculator.

Fig. 2.3 Digital multimeter.

2.6 COMPLEX NUMBERS

Although mathematicians use the variable i to represent $\sqrt{-1}$, electrical engineers use j (as in Section 1.12), since i is reserved for current, from a French word for intensity. Python only accepts j to represent imaginary numbers, and this text follows the EE convention of also using j. A complex number such as $4+3j$ is composed of both real and imaginary parts, and it is often drawn in the complex plane, as shown in Fig. 2.4. There are three common ways of writing complex numbers: rectangular (Cartesian), polar, and complex exponential.

DIGGING DEEPER

Since a positive multiplied by a positive is positive and a negative multiplied by a negative is also positive, it is not possible to find a real number that when multiplied by itself makes a negative number. We therefore must define $\sqrt{-1}$ as something that is not real: we call it j, so that $j^2 = -1$. It is ironic that one of history's great mathematicians, Euler, discovered several important relationships involving complex numbers, including the famous Euler's identity, $e^{j\theta} = \cos\theta + j\sin\theta$, but ultimately dismissed them, saying they were mathematical curiosities of ultimately no practical significance. Now they are used throughout signal processing, communication theory, and control theory.

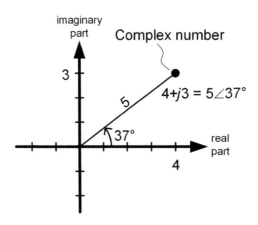

imaginary part

Complex number

3

5

4+j3 = 5∠37°

37°

real part

4

Fig. 2.4 Cartesian and polar representations of complex numbers.

2.6.1 Rectangular Notation

Rectangular notation of complex numbers may be the most familiar to you. The complex number **z** in Fig. 2.4 is written as the sum of its real part and its imaginary part, or $z = 4 + 3j$. Notice that complex variables are bolded. Mathematicians would write the same number as $4 + 3i$. Although electrical engineers place the j in front of its coefficient for emphasis, Python requires that the j be placed after the number without a space. You also must write $1j$ and not j by itself, or you will get an error message.

2.6.2 Polar Notation

The same complex number **z** can also be written as its magnitude, or distance away from the origin, and the angle that a line drawn from the origin to the complex number **z** makes with respect to the positive horizontal axis. This is called polar notation. See Fig. 2.4 for a graphical example of $z = 5\angle 37°$, or $5\angle 0.64$ radians. Note that this represents the same number as $4 + 3j$, expressed in rectangular coordinates.

2.6.3 Complex Exponential

Euler's identity paved the way for a third representation of complex numbers called complex exponential notation, closely related to polar, but which is a function that can be evaluated. Using the above example, $z = 5e^{j0.64}$. Note the similarity to this and the polar form $\mathbf{z} = 5\angle 0.64$, and that numbers in complex exponential form are always written in radians, whereas polar form can be written with the angle either in degrees or in radians.

2.6.4 Entering Complex Numbers in Python

Python uses the rectangular notation to enter and display complex numbers. Thus, to create a complex number such as $4 + 3j$ in Python, type the expression

```
>>> z = 4 + 3j
```

Although implied multiplication is not supported, for the case of the imaginary number, place the number directly in front of the `j` without a space, not behind it. Also note that you cannot use `j` by itself. As a result, `j` must be written as `1j` in Python.

2.6.5 Real and Imaginary Parts

The real part of a complex number `z` can be isolated by importing `NumPy` with the alias `np` and typing

```
np.real(z) or z.real
```

Similarly, the imaginary part of a complex number `z` can be isolated using

```
np.imag(z)  or  z.imag
```

2.6.6 Converting from Rectangular to Polar Form

Polar form is written with a magnitude and angle. Python cannot write the entire number **z** in polar form, but it can extract the magnitude and angle components. The magnitude of a complex number **z** can be found using

```
>>> np.absolute(z)
```

The angle in radians is

```
>>> np.angle(z)
```

To find the angle in degrees use

```
>>> np.angle(z, deg=True)
```

Once the magnitude and angle components are found, the number can be handwritten in polar notation using the format *magnitude* $\angle$ *angle* if the angle is in radians, or *magnitude* $\angle$ *angle*° if the angle is in degrees: for example, $5\angle 37°$.

As an example, to enter the number **w** $= 2 - j6$ in Python and write it on paper in polar form with the angle portion in degrees,

```
>>> w = 2 - 6j
>>> mag = np.absolute(w) # magnitude of w
>>> ang = np.angle(w, deg=True) # angle of w in degrees
>>> print(f"{mag:0.2f} ∠ {ang:0.2f}°")
6.32 ∠ -71.57°
```

2.6.7 Converting from Polar to Rectangular Form

You can use trigonometry and the complex plane diagram, shown in Fig. 2.4, to derive the equations that convert polar to rectangular form, allowing you to enter polar notation numbers in Python.

Given the following complex number in polar form, **z** $=$ mag $\angle$ ang, where the angle is in radians, the real part and imaginary part can be computed using the equations

$$\text{realpart} = \text{mag} \bullet \cos(\text{ang}) \tag{2.1}$$

$$\text{imagpart} = \text{mag} \bullet \sin(\text{ang}) \tag{2.2}$$

If the angle is given in degrees, convert it to radians first. For example, given the polar number **z** $= 5\angle 22°$, find it in rectangular form using

```
>>> realpart = 5 * np.cos(22 * np.pi/180)
>>> imagpart = 5 * np.sin(22 * np.pi/180)

>>> print (f"{realpart:0.2f} + {imagepart:0.2f} j")

4.64 + 1.87 j
```

Once the real and imaginary parts are computed, the complex number can be assembled as described earlier in Section 2.6.4

```
>>> z = realpart + 1j * imagpart
```

2.6.8 Converting to and from Complex Exponential Form

It is possible to convert a complex number in exponential form such as $\mathbf{z} = \text{mag } e^{j\theta}$ with θ in radians into rectangular form simply by entering it into Python. Python always displays complex numbers in rectangular form. For example, $3e^{j\pi}$ can be entered directly as

```
>>> z = 3 * np.exp(1j*np.pi)
```

To convert a number in complex rectangular form to complex exponential form, first convert it to polar form as described previously, and then write the magnitude and angle components in the form $\mathbf{z} = \text{mag } e^{j\theta}$. For example, let $\mathbf{z} = 4 + 3j$. Converted to polar notation (Section 2.6.6) this is $5\angle 0.64$, which in complex exponential notation is then written as $5e^{j0.64}$. In Python, this would be

```
>>> z = 5 * np.exp(0.64j*np.pi)
```

NOTE

Need characters such as $\angle$, π, and Ω in Microsoft Word? Find them all using Insert → Symbol using the Symbol font (important for $\angle$). Want to write them faster? Do this: File → Options → Proofing → Autocorrect → Autocorrect Options → Math Autocorrect → Check the "Use Math Autocorrect outside of Math Regions" option. Then in your document, type \angle to get $\angle$, \pi to get π, and \Omega to get Ω. Note that these commands will only work in the desktop version of MS Word, not the online version.

PRACTICE PROBLEMS

2.5 Write the complex number $z = 5 + 12j$ in polar form with the angle in degrees. ©®

2.6 Write the same complex number $z = 5 + 12j$ in complex exponential form with the angle in radians. ©®

2.7 Use Python to convert $14\angle 45°$ into rectangular form. ©®

2.7 VECTOR MATHEMATICS

Many `NumPy` functions take both scalar and vector arguments, and with one important difference, these work as you would expect: functions that take a single argument, such as `cos()`, operate on each element of the vector independently; operators that take two vectors, like `+`, operate on respective elements of each vector. For instance, if `x = np.array([2, 4, 7])`, then `cos(x)` returns a vector of all three cosines and `x + x` returns the `NumPy array [4, 8, 14]`. Functions that you have used that take vector arguments include `+`, `-`, `sqrt()`, `exp()`, and all the trigonometric functions such as `sin()`.

Vector mathematics using the four basic operations of addition, subtraction, multiplication, and division operates on respective elements. For example,

```
>>> import numpy as np
>>> v = np.array([6, 10])
>>> w = np.array([2, 5])
>>> print(v+w)
[ 8 15]
>>> print(v-w)
[4 5]
>>> print(v*w)
[12 50]
>>> print(v/w)
[3. 2.]
```

All four basic operators work when one argument is a scalar and one is a vector. For example,

```
>>> import numpy as np
>>> x = np.array([2, 4, 9])
>>> print(x-1)
[1 3 8]
```

is the same as

```
>>> import numpy as np
>>> x = np.array([2, 4, 9])
```

```
>>> y = np.array([1, 1, 1])
>>> print(x-y)
[1 3 8]
```

Similarly,

```
>>> import numpy as np
>>> x = np.array([2, 8, 6])
>>> print(x/2)
[1. 4. 3.]
```

gives the same result as

```
>>> import numpy as np
>>> x = np.array([2, 8, 6])
>>> y = np.array([2, 2, 2])
>>> print(x/y)
[1. 4. 3.]
```

PRACTICE PROBLEMS

2.8 Given `x = np.array([4, 2, 6])`, write the Python command to add 5 to each element. ©

2.9 Given `x = np.array([4, 2, 6])`, and `y = np.array([2, 4, 8])`, write the Python command to multiply corresponding elements together. ©

2.10 Calculate $\cos(0)$, $\cos(\pi/2)$, $\cos(\pi)$, $\cos(3\pi/2)$, and $\cos(2\pi)$ using Python and only using `np.pi` once.

2.8 STRINGS

Although strings are not strictly a calculator-like function, Python treats strings as arrays or vectors of characters.

To enter a string, enclose it in either single or double quotation marks. For example

```
s1 = 'Hello, there!' or s1 = "Hello, there!"
```

To concatenate strings, or set/retrieve individual letters or substrings, use the same commands as if working with numerical vectors. For example,

- To retrieve the second letter of `s1` defined above, an "e," type
  ```
  s1[1]
  ```
- To return the first word "Hello" in the string, type
  ```
  s1[0:5]  or  s1[:5]
  ```
- To insert the word "Bonnie" before the exclamation point, use the command
  ```
  s1 = s1[0:12] + 'Bonnie' + s1[12]
  ```
- To delete everything except the word "Hello," type:
  ```
  s1 = s1[0:5]  or  s1 = s1[:5]
  ```

RECALL
Python uses both single and double quotes interchangeably for strings. It makes absolutely no difference to Python. You will see programmers using both conventions.

2.8.1 Converting Between String Representations of Numbers and the Actual Number

One could define a string representing a number, such as

```
string2 = '521'
```

While `string2` looks like a number, it is a string, which is really a vector array of characters. Attempting to perform any numerical operations on `string2` will generate an error message. To convert a string representation of a number into the number, use either `int()`, `float()`, or `complex()`, depending on the numeric class. For example,

```
int('521') + 3
```

now returns the number 524 as expected. Using the `type` command will make this clear. For example,

```
>>> string2 = '521'
>>> print(type(string2))
<class 'str'>
>>> print(type(int(string2)))
<class 'int'>
```

PRACTICE PROBLEMS

2.11   If `strR1 = 'R1 = 47 ohms'`, write the Python command that isolates the number 47 from the string, converts it into a number, and stores it in variable R1. ©

TECH TIP: RESISTORS IN SERIES AND PARALLEL

The most common electronic circuit component is the resistor. Our analogy in the Chapter 1 Tech Tip *Voltage, Current, Charge, and Resistance* following Section 1.10 likened resistors to restrictions in a pipe that limit the flow of water (current) when driven by a constant pressure (voltage). Two resistors of resistance R_1 and R_2 can be connected in series or parallel to create a new equivalent resistance, as shown in Table 2.3.

It is important for EE students to develop their intuition so that the mathematical equations become obvious. Using the water analogy, think of two resistors in series as two restrictions in series, as shown in Table 2.3. A pump (voltage source) must push water (current) through the first, and then through the second, so it follows that the equivalent resistance is the sum of the two individual resistances, or $R_1 + R_2$. The water analogy for two resistors in parallel shows that adding a second resistor increases the number of paths by which water can flow, so it decreases resistance, much like as opening a

TABLE 2.3 Series and parallel resistor combinations and their water analogy

	Two Resistors	Equivalent Resistance	Water Analogy
Series	R_1 ——WW—— R_2 WW	R_1+R_2 WW	R_1 R_2
Parallel	R_1 R_2	$\dfrac{R_1 \cdot R_2}{R_1+R_2}$ WW	R_1 R_2

second adjacent door will decrease resistance to pedestrian traffic flow. The equivalent resistance to two resistors in parallel is

$$\frac{R_1 \cdot R_2}{R_1 + R_2} \tag{2.3}$$

This formula is often memorized as "product over sum" and will always return an equivalent resistance lower than either of the resistances that compose it.

PRACTICE PROBLEMS

2.12 Write the Python command that, given a resistor of 60 Ω and a resistor of 120 Ω, computes the equivalent resistance of the resistors in parallel. Use the variables R_6 and R_7 in your Python command, not the numbers they hold. For example, in series, the Python command is `R6+R7`. ©

2.9 MATRICES

Thus far, the vectors discussed in this chapter have been simple groupings of numbers. For example, a group of three numbers, which might be used to specify a location in three-dimensional space, could be defined as the NumPy array [1, 2, 3]. This section introduces a second category of number groupings, called a *matrix*. A matrix is a two-dimensional array of numbers: for example,

$$\begin{bmatrix} 2 & -4 & 1.7 \\ 42 & \sqrt{2} & 0 \end{bmatrix}$$

Each dimension of the matrix indicates how many rows and columns it has, with the number of rows listed first, followed by the number of columns. Thus, the 2-row × 3-column matrix above is a 2 × 3 matrix. Matrices are used in digital signal processing to hold streams of data, where the data streams can be very large. For example, a 100-second song recorded at a CD quality of 44,000 samples/second can be represented as a 2-column × 4,400,000-row matrix. The columns represent the left and right channels of the signal, and each row represents a sample, that is, a snapshot in time of the voltage to be applied to a speaker.

2.9.1 Special Matrix Dimensions

If a Python NumPy matrix array has a single row, such as 1 x 3 matrix, it is called a *row matrix*. For example, the NumPy array([[1, 2, 3]]) is a two-dimensional matrix array with one row and three columns. Note the extra brackets. Do not confuse this with a one-dimensional NumPy vector array such as we previously discussed. For example, contrast it with the NumPy vector array array([1, 2, 3]). Although both hold the same numbers, a row matrix is a two-dimensional structure that happens to have a single row. A given element still requires both a row and a column index. Compare the indexing used to extract the number 2 from the one-dimensional vector array and the two-dimensional row matrix array in the code

```
>>> import numpy as np
>>> oneDVector = np.array([1, 2, 3])
>>> oneDVector[1]
2
>>> twoDRowMatrix = np.array([[1, 2, 3]])
>>> twoDRowMatrix[0, 1]
2
```

Recall that array indices start at 0, not 1. There will, however, only be one row in the matrix array, with index 0. A Python vector NumPy array is a one-dimensional structure and has neither rows nor columns. A location in a vector array is described using a single index.

Similarly, a *column matrix* has a single column, such as a 3 × 1 matrix. For example, the `numpy` `array([[1], [2], [3]])` is a 3 × 1 column matrix array with three rows and one column. It is still fundamentally a two-dimensional array structure, unlike the one-dimensional `numpy` vector `array([1, 2, 3])`. Notice the indexing needed to extract the value 2 from the column matrix in the code

```
>>> import numpy as np
>>> twoDColMatrix = np.array([[1], [2], [3]])
>>> twoDColMatrix[1, 0]
2
```

Two indices are required, but there will be only one column index, and its value will always be 0. Matrices may have a single row *and* single column in which they hold a single number indexed by row index = 0 and column index = 0, such as the `numpy array([[7]])`. Do not confuse this with a *scalar*, which is zero-dimensional, such as simply `7`, and requires no indices to access its value. Matrices may even be *null*, such as the `numpy array([[]])`, meaning that they are empty and hold nothing.

2.10 CREATING TWO-DIMENSIONAL MATRIX ARRAYS

As with creating vectors, there are several different ways to create matrices:

- Defining components explicitly
- Filling with duplicate values
- Filling with vectors
- Filling with random values

2.10.1 Defining Components Explicitly

To create the two-dimensional matrix $x = \begin{bmatrix} 2 & -3 \\ 1 & 0 \end{bmatrix}$, type the Python commands

```
>>> import numpy as np
>>> x = np.array([[2, -3], [1, 0]])
>>> print(x)
[[ 2 -3]
 [ 1  0]]
```

Compare that with how you earlier defined the one-dimensional vector [2 −3]:

```
>>> import numpy as np
>>> y = np.array([2, -3])
>>> print(y)
[2 -3]
```

Notice that each row of the two-dimensional matrix is defined with a set of enclosing square brackets and separated from the following row by a comma. It is an array of arrays.

2.10.2 Filling with Duplicate Values

The NumPy zeros() and ones() functions can work to create matrices, as well as vectors as described previously. For example, the Python commands

```
>>> import numpy as np
>>> y = np.zeros([2, 4])
>>> print(y)
[[0. 0. 0. 0.]
 [0. 0. 0. 0.]]
```

create a matrix filled with two rows of four columns and are equivalent to typing

```
>>> import numpy as np
>>> y = np.array([[0., 0., 0., 0.],[0., 0., 0., 0.]])
>>> print(y)
[[0. 0. 0. 0.]
 [0. 0. 0. 0.]]
```

Similarly, the Python commands

```
>>> import numpy as np
>>> y = np.ones([3, 2])
>>> print(y)
[[1. 1.]
 [1. 1.]
 [1. 1.]]
```

create a matrix filled with three rows of two columns and is equivalent to typing

```
>>> import numpy as np
>>> y = np.array([[1., 1.],[1., 1.],[1., 1.]])
>>> print(y)
[[1. 1.]
 [1. 1.]
 [1. 1.]]
```

To create a 20 × 30 matrix called `big` filled with 7's, use the fact that Python understands multiplying a matrix by a scalar, like this:

```
>>> import numpy as np
>>> big = 7 * np.ones([20, 30])
```

NOTE

The Python numpy vector array, row matrix array, and column matrix array are all distinctly different. For example, you created a *vector array* of five repeating 1's using `numpy.ones(5)`. This holds five numbers in a one-dimensional array. It has no row or columns; it is just a list of numbers. The row matrix array `numpy.ones([1,5])` is different; it is a two-dimensional array that has one row and five columns. Similarly, the column matrix array `numpy.ones([5,1])` is a two-dimensional array with five rows and one column. The only similarity among these three structures is that they all hold five numbers.

2.10.3 Filling with Vectors

Vectors can be combined to form matrices. Earlier we described how to concatenate the vectors `x = [1 2 3]` and `y = [4 5 6]` into one long vector `z` by using the `numpy concatenate((x,y))` function.

```
>>> import numpy as np
>>> x = np.array([1,2,3])
>>> y = np.array([4,5,6])
>>> z = np.concatenate((x,y))
>>> print(z)
[1 2 3 4 5 6]
```

To create a matrix using vectors, use the `numpy array` function with each vector separated by commas to turn it into a separate row in the created matrix. For example,

```
>>> import numpy as np
>>> x = np.array([1,2,3])
>>> y = np.array([4,5,6])
>>> z = np.array([x,y])
>>> print(z)
[[1 2 3]
 [4 5 6]]
```

2.10.4 Filling with Random Values

Matrices filled with random numbers between 0 and 1 may be created using the `random.rand()` function. For instance, to generate a 3 row × 3 column matrix called `mrand` of three rows and three columns, filled with nine numbers varying between −1 and +1, type the following:

```
>>> import numpy as np
>>> mrand = 2 * np.random.rand(3,3) - 1
>>> print(mrand)
[[-0.9937812 -0.37678119  0.72657737]
 [ 0.0448398  0.53211515  0.28570087]
 [-0.7534636  0.00832234 -0.56896799]]
```

PRACTICE PROBLEMS

2.13 Write the Python commands to generate the following matrices:
1. A matrix with the first row containing −2 and 3, and the second row containing 100 and 0. ©
2. A matrix of 3 rows and 5 columns filled with 0's. ©
3. A column matrix of 10 rows and 1 column filled with 2's. ©

4. Create vectors R1 with eight integers counting up from 1 to 8, and R2 with eight integers counting down from 8 to 1. Write the command to create a 2-row × 8-column matrix with R1 on top and R2 on the bottom. ©

5. Create a 3-row × 2-column matrix of random numbers between 0 and 10. ©

2.11 CHANGING MATRIX VALUES

2.11.1 Accessing and Changing Matrix Values

Matrices are addressed using square brackets with the first index for the row and the second for the column. The row and column index counting always begins with 0. For example, if the matrix

$$m = \begin{bmatrix} -5 & 7 & 2.4 \\ 8 & 0 & 9 \end{bmatrix}$$

is declared as

```
>>> m = np.array([[-5, 7, 2.4],[8, 0, 9]])
```

then

```
>>> x = m[0,1]
```

will set the scalar variable x equal to 7.0. Note once again the zero-based indexing: The first row corresponds to index 0 and the second column corresponds to index 1. The statement

```
>>> m[1,2] = 3.14159
```

replaces the value at the second row and third column to give us the matrix

$$m = \begin{bmatrix} -5 & 7 & 2.4 \\ 8 & 0 & 3.14159 \end{bmatrix}$$

2.11.2 Accessing and Changing Whole Rows and Columns

Much as, if

```
>>> v = np.array([-2, -4, -6, -8, -10])
```

then `v[1:3]` returns `np.array([-4, -6])`, we can address parts of matrices using the colon (:) operator.

Using the example matrix m from the previous section, where

$$m = \begin{bmatrix} -5 & 7 & 2.4 \\ 8 & 0 & 9 \end{bmatrix}$$

`m[0:2,0:2]` returns the square matrix

$$\begin{bmatrix} -5 & 7 \\ 8 & 0 \end{bmatrix}$$

The colon operator is even more powerful than this. It can be used to specify an entire row or column using the shortened form `m[:, :]`. For example, to set x equal to the second row of m, type `x = m[1,:]`

to make `x = [8. 0. 9.]`

To set the first column of m to $\begin{bmatrix} 1 \\ 2 \end{bmatrix}$, type

```
>>> m[:,0:1] = np.array([[1],[2]])
```

Now

$$m = \begin{bmatrix} 1 & 7 & 2.4 \\ 2 & 0 & 9 \end{bmatrix}$$

2.11.3 Removing Matrix Elements

As with vectors, remove elements from matrices using the `NumPy delete ()` method. Note that you can delete any number of whole rows or columns from a matrix. Unlike vectors, however, it does not make sense to delete a single value. If a single value is deleted from a matrix, it will no longer be square or rectangular.

Consider the matrix

$$m = \begin{bmatrix} -5 & 7 & 2.4 \\ 8 & 0 & 9 \end{bmatrix}$$

First create the matrix m; that is,

```
>>> import numpy as np
>>> m = np.array([[-5, 7, 2.4],[8, 0, 9]])
```

To delete rows, use

```
>>> np.delete(array_name, row, 0)
```

In this case, the array_name is m. The third value, 0 in this case, is called the axis. Axis 0 specifies rows. For example,

```
>>> m_del_row0 = np.delete(m, 0, 0)
>>> print(m_del_row0)
```

returns [[8. 0. 9.]].

```
>>> m_del_row1 = np.delete(m, 1, 0)
>>> print(m_del_row1)
```

returns [[-5. 7. 2.4]]. To delete columns, use

```
>>> np.delete(array_name, col, 1)
```

The third value, 1 specifies that columns are to be deleted. For example,

```
>>> m_del_col0 = np.delete(m, 0, 1)
>>> print(m_del_col0)
[[7.   2.4]
 [0.   9. ]]
```

and

```
>>> m_del_col2 = np.delete(m, 2, 1)
>>> print(m_del_col2)
[[-5. 7.]
 [ 8. 0.]]
```

In all the examples, a new matrix is formed and stored in the target variable. To delete rows or columns in the original matrix, use the form

```
>>> m = np.delete(m, 1, 0)
```

The contents of *m* are now [5. 7. 2.4]. A few more examples include

b = np.delete(a,4,1)	delete column 4 and store the result in b
b = np.delete(b,[2,3],0)	delete rows 2 and 3 of b
d = np.delete(c,[0,2],0)	delete row 0 and 2 and store the result in d
d = np.delete(d,[0,3,4],1)	delete columns 0, 3, and 4 of d

PRACTICE PROBLEMS

2.14 Using the following matrix, write the Python commands to solve the following problems. First, define matrix m as

$$m = \begin{bmatrix} 1 & 2 \\ 3 & 4 \end{bmatrix}$$

1. Create a variable x and set it equal to the second row, second column value of m. ©
2. Set the second row, first column value of m to 0. ©
3. Replace the first column of m with random numbers created using random.rand(). ©
4. Create a vector v equal to the last column of m. ©
5. Remove the entire top row of m. ©

PRO TIP: BUYING A CALCULATOR

Electrical engineers in industry have their own offices and typically perform computations using company software. But for college students, one of the most important tools is the portable calculator. Unlike most majors, ECE students must have calculators capable of handling complex matrices to solve problems involving circuits driven by sinusoidal sources; the TI-86 and below will not do this, including the TI-84. Do not be misled into purchasing an expensive instrument with many specialized engineering problems built into it; these are rarely useful. A TI-89, TI-92, or higher, or the newer TI-Nspire series, all work well. The Nspire's document-centric nature is unique; some students prefer it, but many see it as an abstraction that separates them from simply solving equations. Before purchasing, check with your university to see if they have any restrictions on calculator use in exams.

2.12 WORKING WITH MATRICES

2.12.1 Matrix Array Shape

Just as the `len()` command can be used to determine how many elements are in a vector, the `shape` command returns how many rows and columns are in a matrix. For example, if

```
>>> x = np.array([[4, 3],[2, -4], [5, 7]])
```

that is, $x = \begin{bmatrix} 4 & 3 \\ 2 & -4 \\ 5 & 7 \end{bmatrix}$, then the matrix dimensions can be found using

```
rows, cols = np.shape(x) or rows, cols = x.shape
```

This sets the variable `rows` to 3 and `cols` to 2. Note how `shape` sets two variables at the same time. When `shape` is used on a vector, it gives the length, followed by a comma. In the following examples note the subtle difference between the way Python represents an n-dimensional vector, a $1 \times n$ matrix, and an $n \times 1$ matrix.

```
>>> import numpy as np
>>> a = np.array([10, 20, 30, 40, 50]) # n=5 dimensional vector
>>> print(a.shape)
(5,)
>>> b = np.array([[10, 20, 30, 40, 50]]) # 1x5 array
>>> print(b.shape)
(1,5)
>>> c = np.array([[10], [20], [30], [40], [50]]) # 5x1 array
>>> print(c.shape)
(5,1)
```

2.12.2 Transpose

Transposing a real matrix interchanges rows and columns. For example,

$$\begin{bmatrix} 4 & 3 \\ 2 & -4 \\ 5 & 7 \end{bmatrix} \text{ becomes } \begin{bmatrix} 4 & 2 & 5 \\ 3 & -4 & 7 \end{bmatrix} \text{ when transposed.}$$

The Python command to do this is the `NumPy transpose ()` function. For example,

```
>>> import numpy as np
>>> x = np.array([[4, 3],[2, -4], [5, 7]])
>>> print(x)
[[ 4  3]
 [ 2 -4]
 [ 5  7]]
>>> print(np.transpose(x))
[[ 4  2  5]
 [ 3 -4  7]]
```

This can also be used to change a column matrix to a row matrix and vice versa. For example, defining the column matrix

```
>>> Resistors = np.array([[10],[11],[12],[13],[14]])
```

is the same as defining the transpose of the row matrix as

```
>>> Resistors = np.transpose(np.array([[10, 11, 12, 13, 14]]))
```

and both define the column matrix $\begin{bmatrix} 10 \\ 11 \\ 12 \\ 13 \\ 14 \end{bmatrix}$.

2.12.3 Matrix Mathematics

Matrix mathematics using the basic operations of addition, subtraction, multiplication, division, and exponentiation operate on the respective elements. For example,

$$\begin{bmatrix} 1 & 5 \\ 2 & 3 \end{bmatrix} + \begin{bmatrix} -1 & 6 \\ 0 & 1 \end{bmatrix} = \begin{bmatrix} 0 & 11 \\ 2 & 4 \end{bmatrix}$$

Addition, subtraction, multiplication, division, and exponentiation also work with a matrix and a scalar:

$$\begin{bmatrix} 1 & 5 \\ 2 & 3 \end{bmatrix} + 2 = \begin{bmatrix} 3 & 7 \\ 4 & 5 \end{bmatrix}$$

PRACTICE PROBLEMS

2.15 What single Python command will allow you to evaluate the expression

$$y = \frac{\cos\left(\dfrac{t}{2} + \dfrac{\pi}{4}\right)}{t + 1} \tag{2.4}$$

given the vector array of *t* values `t=np.linspace(0,5,10)`? ©®

TECH TIP: MESH AND NODAL ANALYSIS METHODS

Techniques for finding currents and voltages in complex circuits with many resistors include node voltage analysis and mesh current analysis. These methods are usually taught in circuit analysis courses, but at their core, they are a systematic way of applying Ohm's Law $V = IR$ repeatedly over circuit fragments. They create systems of equations that can be solved simultaneously to find every voltage (for nodal voltage analysis) or current (for mesh current analysis) in the circuit. Python can solve these simultaneous equations, which sometimes include more than 100 unknowns, in less than a second.

The diagram shown in Fig. 2.5a gives a circuit, its voltage definitions, and the set of equations generated by the node voltage analysis method for its voltages. Figure 2.5b shows the same circuit with its currents defined, as well as the set of equations derived using the mesh current analysis method to find these currents. Beginning EE students should know the names of these methods, and you will learn how to apply them in your first circuit analysis course.

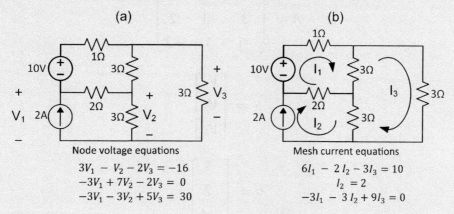

(a)

(b)

Node voltage equations

$$3V_1 - V_2 - 2V_3 = -16$$
$$-3V_1 + 7V_2 - 2V_3 = 0$$
$$-3V_1 - 3V_2 + 5V_3 = 30$$

Mesh current equations

$$6I_1 - 2I_2 - 3I_3 = 10$$
$$I_2 = 2$$
$$-3I_1 - 3I_2 + 9I_3 = 0$$

Fig. 2.5 Node and mesh circuit analysis. (a) The node voltage analysis method produces a set of equations that can be solved for the voltages. (b) The same circuit can be analyzed using a mesh current analysis that solves for the currents.

2.13 SOLVING SIMULTANEOUS EQUATIONS USING MATRICES

Since nodal analysis and mesh analysis are very common methods of circuit analysis and yield sets of linear real or complex simultaneous equations, solving simultaneous equations is a staple of the electrical engineering curriculum. In Python, this is accomplished as shown in the following example.

Solve for the various voltages V_1, V_2, and V_3 in the following set of equations:

```
V₁ + 2 V₂ + V₃ = 5
3 V₁ - V₂ + 2 V₃ = 2
-V₁ + V₂ - 2 V₃ = -4
```

Students with a background in matrix algebra will recognize this as a single matrix equation of the form

$$A x = b \tag{2.5}$$

where

$$A = \begin{bmatrix} 1 & 2 & 1 \\ 3 & -1 & 2 \\ -1 & 1 & -2 \end{bmatrix}$$

$$x = \begin{bmatrix} V_1 \\ V_2 \\ V_3 \end{bmatrix}$$

$$b = \begin{bmatrix} 5 \\ 2 \\ -4 \end{bmatrix}$$

NOTE

The matrix A must be square, and the inverse must exist.

Whether or not one has a background in matrix algebra, they can be solved using the following steps. First, define the matrix *A* with the coefficients of the unknown variables, and define a single column matrix *b* with the values to the right of the equals sign, as follows:

```
>>> import numpy as np
>>> A = np.array([[1, 2, 1], [3, -1, 2], [-1, 1, -2]])
>>> b = np.array([[5],[2],[-4]])
```

To find the solution, use the NumPy linalg subpackage function called solve:

```
x = np.linalg.solve(A, b)
```

When using nodal and mesh methods to analyze the response of circuits containing capacitors and inductors, you will solve sets of complex matrices. There is no difference as far as Python is concerned; the *A* matrix will have complex values, as for this example solving for complex currents:

```
>>> A = np.array([[1+1j, 2-2j],[-3+1j, 3]])
>>> b = np.array([[1j],[2-1j]])
>>> I = np.linalg.solve(A, b)
```

On a modern computer, Python can solve a set of 100 equations with 100 unknowns in about 1 ms—less time than it takes to depress the Enter key fully.

DIGGING DEEPER

Matrices can do more than solve simultaneous equations for circuit analysis. They are essential to manipulating video, rotating photos, providing equalization to audio streams, finding specific frequencies in signals, and controlling servo motors, and are the discrete mathematics counterpart of integration and differentiation.

PRACTICE PROBLEMS

2.16 Use Python to solve the following set of mesh equations for a circuit to find I_1, I_2, and I_3: ©®

$$2 I_1 + 6 I_2 - 3 I_3 = 3$$
$$4 I_1 - I_2 + I_3 = 25$$
$$I_1 + 2 I_2 - I_3 = 8$$

TECH TIP: VOLTAGE DIVIDERS

One of the most common circuits is the voltage divider. It is composed of two resistors in series so that a voltage applied across both is split, with part being dropped by the first resistor and the remainder by the second as shown in Fig. 2.6.

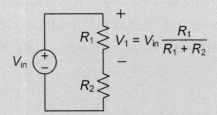

Fig. 2.6 Voltage divider circuit.

As with all new ECE formulas, take a moment to analyze it:

- The voltage drop across R_1 is proportional to the applied voltage V_{in}, which makes intuitive sense; if one doubles the applied voltage, the voltage drop across each of the resistors should double.
- The amount of voltage dropped by R_1 is proportional to the ratio of R_1 to the total circuit resistance $R_1 + R_2$. Since the entire applied voltage must be dropped by the two resistors, it is intuitive that it is split according to each resistor's relative contribution to the total resistance.
- By the same reasoning, the voltage across R_2 is

$$V_2 = V_{in}\frac{R_2}{R_1 + R_2} \tag{2.6}$$

This means the sum of the voltage drops across both resistors is

$$V_{in}\frac{R_1}{R_1 + R_2} + V_{in}\frac{R_2}{R_1 + R_2} \tag{2.7}$$

which simplifies to V_{in}, the total applied voltage, just as we expect.

PRACTICE PROBLEMS

2.17 Using the circuit in the previous Tech Tip with $R_1 = 15\ \Omega$ and $R_2 = 6\ \Omega$, find the V_{in} needed to make $V_1 = 19$ V, using Python and algebra.

COMMAND REVIEW

NumPy package import

`import numpy as np`   Required for commands using `numpy` alias `np`.

Built-in Python Data Types

`Numeric Types`
  `int`   Integer, e.g., `7`, `15`, `52349`.
  `float`   Floating point, e.g., `3.14`, `2.`, `7563.214`.
  `complex`   Complex, e.g., `7+1j`. Note that `7+j` or `7+j1` are incorrect formats.
`Boolean Type`   `True, False`.
`NoneType`   Used to represent type `None` or no value.
`str`   String type, e.g., `'text'` or `'Name'`.
`list`   Ordered collection of mutable objects, e.g., `[1,5]` or `["O","T"]`.
`tuple`   Ordered collection of immutable objects, e.g., `(1, 2, 3)`.
`set`   Unordered collection mutable objects, e.g., `{1, 6, 5}`.
`dictionary`   Set of key value pairs, e.g., `{1:"O", 2:"P"}`.
`type(x)`   Returns object type of `x`, e.g., `type(7)` gives `<class 'int'>`.

Rounding Numbers

`np.round(n)`   Rounds to the nearest integer, so [−3.6, 5.7] becomes [−4, 6].
`np.fix(x)`   Rounds toward zero. Returns the integer part of the number.
`np.floor(x)`   Round down. Outputs the greatest integer less than or equal to `x`.
`np.ceil(x)`   Round up. Outputs the smallest integer greater than or equal to `x`.
`np.modf(x)`   Outputs the fractional and whole parts of a number.

Creating Vector Arrays

`np.array([])`   Creates a vector array by explicitly listing the elements.
`np.zeros(n)`   Creates a vector array of _n_ zeros.
`np.ones(n)`   Creates a vector array of _n_ ones.
`np.arange(a,b,s)`   Creates a linearly spaced vector with start _a_, stop _b_ − _s_, and step _s_.
`np.arange(a,b)`   Creates a linearly spaced vector with start _a_, stop _b_ − 1, and step 1.
`np.arange(b)`   Creates a linearly spaced vector with start 0, stop _b_ − 1, and step 1.
`np.linspace(a,b,n)`   Creates a linearly spaced vector with start _a_, stop _b_, and _n_ elements.
`np.logspace(a,b,n)`   Creates a log spaced vector with start 10^a, stop 10^b, and _n_ elements.
`np.random.rand(1,10)`   Creates a vector _v_ filled with 10 random float numbers between 0 and 1.
`np.arange(a,b)`   Creates a vector beginning at _a_ and ending at _b_ − 1 incrementing by 1.

Vector Array Operations

`len(v)`   Returns the number of elements in a vector array.
`v[i] = np.pi`   Replace the vector array value at index _i_. Remember indices start at 0.
`v[start:stop:step]`   Return a range of vector array values from start, to stop 1, in steps.
`v[start:stop]`   Returns a range of vector array values from start, to stop 1, in unit steps.
`v[-3:-1]`   Use the negative sign to index from the end.
`w = np.delete(v, i)`   Delete the element at index `i` from `v`.

| `np.concatenate(x, [7, 8], y)` | Concatenate vector arrays. |
| `v = np.insert(v, i, x)` | Insert element `x` at index `i` into `v`. |

Complex Number Operations

`np.real(z)` or `z.real`	Return the real part of a complex number.
`np.imag(z)` or `z.imag`	Return the imag part of a complex number.
`np.absolute(z)`	Return the magnitude part of a complex number.
`np.angle(z)`	Return the angle in radians of a complex number polar representation.
`np.angle(z, deg=True)`	Return the angle in degrees of a complex number polar representation.
`np.angle(z)*180/np.pi`	Return the angle in degrees of a complex number polar representation.

Strings

`v1 = 'Hello'`	Defines a string using single quotes.
`v2 = "Hello"`	Defines a string using double quotes.
`char1 = v1[i]`	Retrieves the *i*th character of the string. Remember, start at 0.
`str3 = str1 + str2`	Concatenates two or more strings using the + operator.
`str4[0:3] = ''`	Deletes a range of the string. Deletes from character 0 to character 2.
`x = int('45')`	Converts a string to its corresponding integer.
`y = float('45.4')`	Converts a string to its corresponding floating point value.
`z = complex('1+2j')`	Converts a string to its corresponding complex value.

Creating Matrix Arrays

| `np.zeros(r,c)` | Creates a matrix of zeros with `r` rows and `c` columns. |
| `np.ones(r,c)` | Creates a matrix of ones with `r` rows and `c` columns. |

Array Operations

`[r, c] = np.shape(x)`	Returns the rows to `r` and columns to `c` of an array `x`.
`[r, c] = x.shape`	Returns the rows to `r` and columns to `c` of an array `x`.
`np.transpose(y)`	Returns the transposition of array `y`.
`+ - * / **`	Works between scalars, scalars and arrays, or arrays componentwise.

Solving Linear Equations Ax = b

| `x = np.linalg.solve(A, b)` | Solves the set of *n* simultaneous equation where `A` is an $n \times n$ matrix of coefficients and `b` is an $n \times 1$ column matrix. |

LAB PROBLEMS

©=Write only the Python command(s).

®=Write only the Python result.

Solutions to all starred (*) problems are available on the Elsevier website (refer to page xix for the link).

2.1* Resistors are available in certain predefined sizes, so you cannot expect to find one with a value of π Ω, for example. Write the Python code that creates the vector $v1$ of all *standard* 5% resistor values from 1 Ω through 9.1 Ω. The list of standard values is described in the Tech Tip *Measuring Resistance* just preceding Section 2.5. ©®

2.2 Create vector $v2$ by modifying $v1$, from lab problem 2.1, so that it prints all standard 5% resistor values from 1 Ω through and including 1 MΩ, which is the same as 1,000,000 Ω. You will use these vectors again, so save both $v1$ and $v2$ in the Jupyter Notebook resistors.ipynb. Hint: How would you scale $v1$ to make it 10 times bigger? How would you then concatenate $v1$ and that scaled $v1$? Make sure the last value is 1 MΩ, not 910,000 Ω. ©

2.3 What is the Python code that removes the third through seventh values of a vector v? Hint: First generate a vector *indx* of integers 2 through 6 that represents the indices to be removed. Then use the `np.delete` command to delete *indx's* indices in v. ©

2.4* What single `NumPy` method removes the first, second, and fifth columns from a matrix c? ©

2.5 What Python command replaces the first column of any matrix m with all zeros? ©

2.6 What Python command replaces the second row of a matrix m with the vector v? Assume the number of columns in m is the same as in v. For instance, if `m = np.array([[1, 2], [3, 4]])` and `v = np.array([9, 9])`, this will create the matrix `[[1, 2], [9, 9]]`. The command should work with any compatibly-sized m and v. ©

2.7 If you are given a vector x, write the Python code to
   a) extract the first five numbers; ©
   b) delete the last five numbers. ©

2.8* Write the Python code that, given a predefined vector *v* of length 5, returns only the values with even indices. For example, if `v = np.array([1, 5, 3, -3, 0])`, it will return `np.array([1, 3, 0])`. Recall that `NumPy array` indices start at 0, not 1. ©

2.9 Write the Python code that, given a predefined vector *v* of any length, returns only the values with even indices. This is similar to lab problem 2.8 but harder. The code must be the same regardless of the vector length. For example, if `v = np.array([1, 5, 3, -3, 0])`, it will return `[1, 3, 0]`, but if `v = np.array([1, 5, 3])`, it will return `[1, 3]`. This reduction in data length is called *downsampling* when applied to digital signal processing. ©

2.10 Write the Python code needed to remove all even-indexed rows from a matrix *m*, where *m* could be of any size. For instance, in a 6 × 6 square matrix, the command would remove the zeroth, second, fourth, and sixth rows, but in a 2 × 2 matrix, it would only remove the zeroth row. ©

2.11* Make a vector `v3` of the alternating digits [1 −1 1 −1 ...] that is $2^{10} = 1024$ values long. Hint: Do not specify each value independently since that would take too long. Instead, think about how to use the fact that $(-1)^0 = 1, (-1)^1 = -1, (-1)^2 = 1, (-1)^3 = -1$, etc. ©

2.12 Given a vector `x` of any length, what Python code creates a vector `y` consisting of every odd-indexed value in `x`? For example, if `x = np.array([5, 1, 2, 5])`, then `y = np.array([1, 5])`, and if `x = np.array([1, 3])`, then `y = np.array([3])`. ©

2.13 Write the Python code that takes a given vector `g` of any length, and from it, produces its flipped vector `f`. For example, if `g = np.array([1, 2])`, then `f = np.array([2, 1])`, and if `g = np.array([1, 2, 3, 4])`, then `f = np.array([4, 3, 2, 1])`. Hint: Vector increments can be negative. ©

2.14 Given a vector `x` of arbitrary length, write the Python code that creates the vector `y` that doubles the `x` vector length, duplicating each value. For instance, if `x = np.array([2, -1])`, then `np.array([2, 2, -1, -1])` is returned, and if `x = np.array([2, -1, 3])`, then `np.array([2, 2, -1, -1, 3, 3])` is returned. Begin with the code

```
import numpy as np
x = np.array([2, -1])
y = np.zeros(2*len(x))
y[0:-1:2] = x
```

2.29   A mesh analysis of a circuit yields the set of equations

$$2 I_1 - I_2 + I_3 + 6 I_4 = 20$$

$$-I_1 + 3 I_2 = -12$$

$$I_3 + I_4 = 1$$

$$2 I_1 - I_3 = 12$$

Use Python to solve them for I_1, I_2, I_3, and I_4. Hint: If an equation does not mention a variable, there is 0 times that variable. Also, due to limits of machine precision, sometimes an answer that should be exactly an integer (e.g., 0 or 1) may be solved by Python to be slightly different (e.g., 0.0000000000001, or 0.9999999999998). In these cases, you should check to see if rounding it solves the problem, and then use the rounded result. ©®

2.30   A node voltage analysis of a circuit yields the equations

$$3 V_1 - 2 V_2 + V_3 = 0$$

$$V_1 + V_2 + 1 = 0$$

$$V_1 - V_2 + V_3 - 2 = 0$$

Use Python's matrix-solving capabilities to find the three voltages. Hint: Remember to move the constants to the right side. ©®

2.31* A nodal analysis of a circuit containing inductors and capacitors yields the set of complex equations

$$(1 + 2j) V_1 + 3j V_2 = -1 + 12j$$

$$2 V_1 - 4 V_2 = -16 + 34j$$

Use Python to find the complex values V_1 and V_2 and write their values in rectangular form. ®

2.32   A mesh current analysis of a circuit yields the set of complex equations

$$(4-j) I_1 + I_2 + 4-j = 0$$

$$7j I_1 - 3 I_2 = 0$$

Use Python's matrix-solving capabilities to find the two currents in polar form, with the angle in degrees. Hint: Remember to move constants to the right side. ©®

PLOTTING IN PYTHON

3.1 OBJECTIVES

After completing this chapter, you will be able to use Python to do the following:

- Create line plots, similar to an oscilloscope display
- Create multiple line plots on the same axes, similar to a multiple channel oscilloscope
- Create scatterplots of data pairs, such as calculated voltage as a function of measured voltage
- Create bar plots to show grouped data
- Create linear or logarithmically spaced axes
- Create figures with multiple sets of axes subplots
- Add text labels to plots, including Greek symbols
- Create three-dimensional plots
- Save plots in files so that they can be inserted into reports

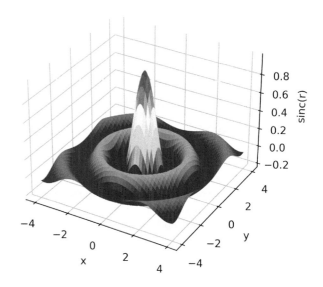

3.2 TYPES OF PLOTS

Electrical engineers use a wide variety of plots to show data. The most common types, illustrated in Fig. 3.1, are line, scatter, and bar plots. The line plot in the top panel is used to show continuously varying data, like an oscilloscope displays a time-varying voltage. The scatterplot is used to show discretely measured data. For example, in a production lot of inductors, six may be sampled and their measured inductance displayed in a scatterplot (panel b). If there is too much data to display using a scatterplot, for instance, if 10,000 resistors were sampled, the data may be grouped into bins and displayed as a bar plot (panel c).

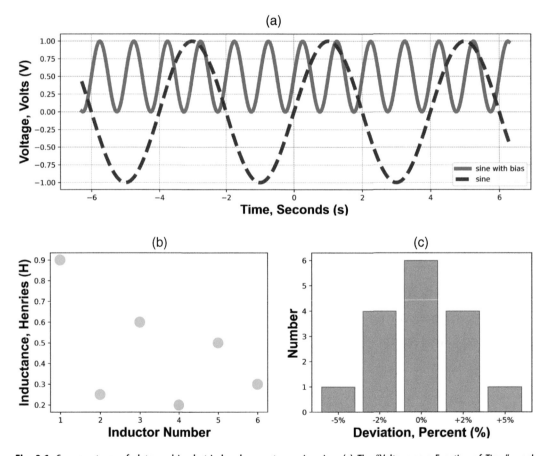

Fig. 3.1 Common types of plots used in electrical and computer engineering. (a) The "Voltage as a Function of Time" graph shows two differently formatted line plots on the same axes. (b) The "Inductance" plot is an example of a scatterplot. (c) A bar chart provides a visual description of the number of samples falling within a specific range. In this chapter, you will learn how to create all these plot types.

3.2.1 Line Plots

The line plot is the most common type of graph. It requires two vector arrays of equal length; one array specifies the horizontal coordinates of each point while the other specifies the corresponding vertical points.

Fig. 3.2, for example, plots the function

$$y(t) = \frac{t}{2}\cos(t), 0 \le t \le 5 \tag{3.1}$$

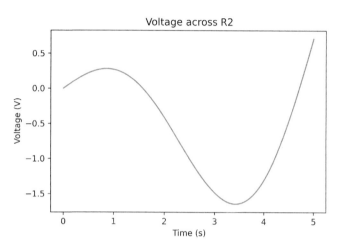

Fig. 3.2 Plot of $y(t) = t/2\cos(t)$ over the interval $0 \le t \le 5$. The vertical axis dependent variable y is plotted as a function of the horizontal axis independent variable t. Note the use of a title and axis labels, which are a hallmark of any professionally created figure.

In high school mathematics classes, horizontal distances are often called x and vertical distances y. In electrical engineering, the horizontal axis is often t for time, and the vertical axis is a voltage $v(t)$, a current $i(t)$, or a generic signal, such as $x(t)$ or $y(t)$. Since the horizontal variable is the input to the function and the output depends on it, the horizontal variable is often called the *independent* variable, and the vertical variable the *dependent* variable.

To create the plot of Fig. 3.2:

1. Use the `import numpy as np` statement to import the NumPy package with the alias `np` to create `arrays` and perform scientific calculations. Similarly, `import` the `Matplotlib pyplot` module with the alias `plt` to create plots. If you are using a Jupyter Notebook, include the magic function `%matplotlib inline` to display your plot directly below the code cell that produced it. Do not use the `%matplotlib inline` statement with the Python interpreter. These statements are usually grouped at the beginning of a program as

```
import numpy as np
import matplotlib.pyplot as plt
%matplotlib inline
```

2. Create the horizontal vector `t` as a `NumPy array` with enough points to give a smooth plot, and create the vertical vector `y` corresponding to the desired function.

```
t = np.linspace(0.0, 5.0, 1000)
y = t / 2.0 * np.cos(t)
```

RECALL

The commands +, −, *, and / work between a scalar and a vector array. Between two vector arrays, these commands act on corresponding elements of each array element.

3. Use the `subplots()` method to create a new default figure with one axis. This command returns two handles: one for the entire figure, called `fig`, and a second for the axis/axes, called `ax`. This provides more flexibility than the simpler method described earlier in Section 1.18 because the axes handle lets us modify the axes later by adding titles, axis labels, and legends.

```
fig, ax = plt.subplots()
```

4. Plot the vector `y` as a function of the vector `t` onto the created axes `ax` using

```
ax.plot(t, y)
```

5. Always title your plot and label your horizontal and vertical axes. This is critically important! Untitled plots of unlabeled variables mean nothing in a document and will lead to confusion. Python makes it simple to set the plot title and axis labels:

```
ax.set_title('Voltage across R2')
ax.set_xlabel('Time (s)')
ax.set_ylabel('Voltage (V)')
```

6. If you are using the Python interpreter, you will need to show the plot using the command

```
plt.show() # Needed with Python interpreter, not in Jupyter
```

The complete Python programs using both the Python interpreter and a Jupyter Notebook code cell are summarized in Fig. 3.3. With a Jupyter Notebook you may want to use a semicolon at the end of the last statement to suppress unwanted output.

```
Python Interpreter                                          >>>
>>> # Python interpreter shell
>>> # Plot y = t/2.0 * cos(t) over the interval 0.0 <= t <= 5.0
>>>
>>> import numpy as np
>>> import matplotlib.pyplot as plt
>>>
>>> t = np.linspace(0.0, 5.0, 1000)
>>> y = t/2.0 * np.cos(t)
>>>
>>> fig, ax = plt.subplots()
>>> ax.plot(t,y)
>>>
>>> ax.set_title('Voltage across R2')
>>> ax.set_xlabel('Time (s)')
>>> ax.set_ylabel('Voltage (V)')
>>>
>>> plt.show() # Needed with a Python interpreter shell
```

```
Jupyter Notebook                                          .ipynb
 1  # Jupyter notebook code cell
 2  # Plot y = t/2.0 * cos(t) over the interval 0.0 <= t <= 5.0
 3
 4  import numpy as np
 5  import matplotlib.pyplot as plt
 6  %matplotlib inline
 7
 8  t = np.linspace(0.0, 5.0, 1000)
 9  y = t / 2.0 * np.cos(t)
10
11  fig, ax = plt.subplots()
12  ax.plot(t,y)
13
14  ax.set_title('Voltage across R2')
15  ax.set_xlabel('Time (s)')
16  ax.set_ylabel('Voltage (V)'); # ; suppresses unwanted output
```

Fig. 3.3 Python interpreter and Jupyter Notebook code used to plot $y(t) = t/2 \cos(t)$ over the interval $0 \le t \le 5$. Note that the Python interpreter needs the `plt.show()` statement to display the plot. When using a Jupyter Notebook, include the `%matplotlib inline` statement to display the plot below the code cell that generated the plot. Use a semicolon (`;`) at the end of the last statement of a code cell to suppress output generated by the statement that is not needed.

 NOTE
Use the `plt.show()` statement to display the plot with a Python interpreter. With a Jupyter Notebook, include the `%matplotlib inline` statement to display the plot below the code cell. Use a semicolon (`;`) to suppress unwanted output produced by the last statement of a code cell.

3.2.2 Selecting Plot Line Colors

The Python `Matplotlib pyplot` module will select blue for plot lines by default. To select a different color, such as plotting the previous example in red, type

```
ax.plot(t,y,'r')
```

where `'r'` stands for red.

There are eight predefined colors with single-letter shortcut abbreviations, summarized in Fig. 3.4.

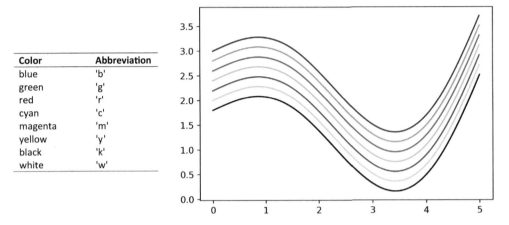

Color	Abbreviation
blue	'b'
green	'g'
red	'r'
cyan	'c'
magenta	'm'
yellow	'y'
black	'k'
white	'w'

Fig. 3.4 Plot line colors. Including the single-letter shortcut abbreviation in the `ax.plot()` statement sets the plot line color.

DIGGING DEEPER

If you need to embed Python plots in Microsoft Word, select the plot window you wish to copy, right-click, choose Copy Image from the drop-down menu, and then paste it into the Word document. This method will copy the figure at the screen resolution of your display, which typically prints with a slight blur. The end of this chapter explains how to save publication-quality images in files for importing into Word and other packages.

3.2.3 Selecting Plot Linestyles

By default, `Matplotlib pyplot` uses solid lines to draw plots. However, other linestyles can be specified, such as dotted, dashed, and dash—dot. The left panel of Fig. 3.5 summarizes the common linestyles used for plotting. To create a dotted line as shown in Fig. 3.5, type

```
ax.plot(t,y,':')
```

This command can be combined with the line color command by typing

```
ax.plot(t,y,'r:')
```

This creates a red dotted line. Notice how the two commands are combined with a single set of quotation marks.

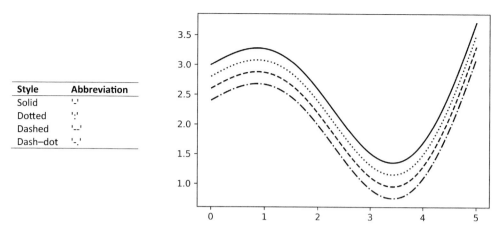

Style	Abbreviation
Solid	'-'
Dotted	':'
Dashed	'--'
Dash–dot	'-.'

Fig. 3.5 Plot linestyles. To set the linestyle, use the abbreviations in the `ax.plot()` command.

3.2.4 Selecting Plot Line Thicknesses

The line width is set to 1 by default. To make it five times thicker, for example, choose a line width of 5 using the command

```
ax.plot(t, y, linewidth=5)
```

Examples of plots with line widths ranging from 1 to 15 are shown in Fig. 3.6.

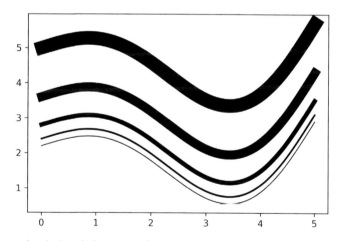

Fig. 3.6 Plot line thicknesses. To set the plot line thickness to 7, for example, for a black solid line, `'k-'`, set the `linewidth` equal to n using `ax.plot(t, y, 'k-', linewidth=7)`.

Similarly, if you need a color other than one of the eight predefined colors that `Matplotlib` provides, use the `color` keyword, followed by the color specification in [red, green, blue] coordinates. This is a vector of three numbers, with each number varying between 0 and 1, that define how much red, blue, and green are present, respectively. This color specification system is common in computer graphics. For example, orange is made with a lot of red, mixed with a medium amount of green, and no blue. The RGB coordinates [1, 0.5, 0] therefore yield an orange hue, which can be plotted using

```
ax.plot(t,y,color=[1, 0.5, 0])
```

3.2.5 Combining Plot Commands

The arguments of `plot()` must be grouped into two parts. The first is the *x, y* data, followed by an optional single-character color and an optional one- or two-character linestyle. Any color and linestyle characters must be grouped together and enclosed in a single set of quotation marks, that is, `'r-'` rather than `'r','-'`. Commands that include the equals sign, such as `linewidth=10`, must come at the end. For example, to draw the thick red dotted line shown in Fig. 3.7, use the command

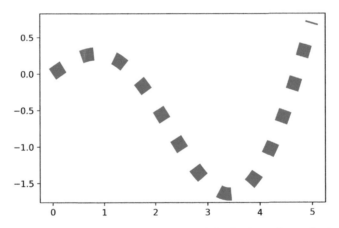

Fig. 3.7 Combining plot commands. Group the plot arguments into two parts, with the first being the *x* and *y* data, followed by an optional single-character color and an optional one- or two-character linestyle.

PRACTICE PROBLEMS

3.1 Plot cos(*t*) from $-2\pi \le t \le 2\pi$. ©®

3.2 Repeat the above using a dotted green line, with a width 10 times thicker than the default. ©®

3.2.6 Axis Labels and Titles

Most plots will need labels for each axis and a title. This can be done after the plot is created using the `set_xlabel()`, `set_ylabel()`, and `set_title()` methods, as shown in Fig. 3.8. This program plots the first 5 seconds of the decaying voltage exponential shown in Fig. 3.9.

```
Python Interpreter                                              >>>
>>> # Decaying exponential plot with labels
>>>
>>> import numpy as np
>>> import matplotlib.pyplot as plt
>>>
>>> t = np.linspace(0.0, 5.0, 100)
>>> y = np.exp(-t)
>>>
>>> fig, ax = plt.subplots()
>>> ax.plot(t,y)
>>> ax.set_title('Decaying Exponential')
>>> ax.set_xlabel('Time, Seconds (s)')
>>> ax.set_ylabel('Amplitude, Volts (V)')
>>> plt.show()
```

```
Jupyter Notebook                                            .ipynb
 1  # Decaying exponential plot with labels
 2
 3  import numpy as np
 4  import matplotlib.pyplot as plt
 5  %matplotlib inline
 6
 7  t = np.linspace(0.0, 5.0, 100)
 8  y = np.exp(-t)
 9
10  fig, ax = plt.subplots()
11  ax.plot(t,y)
12  ax.set_title('Decaying Exponential')
13  ax.set_xlabel('Time, Seconds (s)')
14  ax.set_ylabel('Amplitude, Volts (V)'); #Suppress last line output
```

Fig. 3.8 Python code for a decaying exponential plot with labels. The `ax.set_title()`, `ax.set_xlabel()`, and `ax.set_ylabel()` methods produce a title, *x* label, and *y* label for the plot.

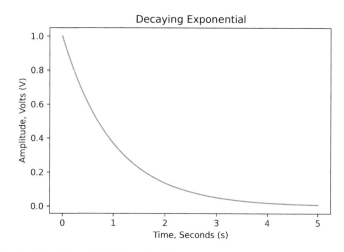

Fig. 3.9 Decaying exponential plot. Set the axis labels and title using `ax.set_xlabel()`, `ax.set_ylabel()`, and `ax.set_title()`, respectively.

3.2.7 Axis Limits

`Matplotlib` usually does a good job of choosing vertical and horizontal axis limits, but not always. For example, to make a right triangle, draw three lines connecting the four points {0,0}, {1,0}, {0,1}, and {0,0} as shown in Fig. 3.10. The last point is the same as the first to close the triangle. The Python code to generate this follows.

```python
import numpy as np
import matplotlib.pyplot as plt

# Use %matplotlib inline with Jupyter Notebook
%matplotlib inline

# points describe lower left, lower right, top, back to start point
x = np.array([0, 1, 0, 0])
y = np.array([0, 0, 1, 0])

fig, ax = plt.subplots()
ax.plot(x,y)
ax.set_title('Plot shapes using vertices')
ax.set_xlabel('x values')
ax.set_ylabel('y values');
plt.show() # Needed with Python interpreter but not a Jupyter Notebook
```

To choose your own axis limits, use the command

```python
ax.set(xlim=(xmin, xmax), ylim=(ymin, ymax))
```

For example, to center the plot in Fig. 3.10, choose horizontal and vertical axis limits that range from −1 to 2:

```python
ax.set(xlim=(-1, 2), ylim=(-1, 2))
```

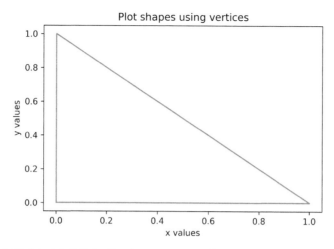

Fig. 3.10 Axis default limits. Use the `ax.set()` function to change the axis limits.

This command will result in a centered plot with a larger border, as shown in Fig. 3.11.

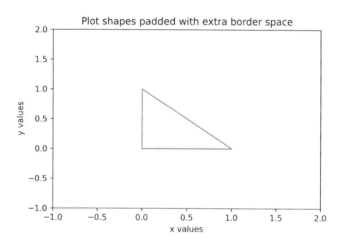

Fig. 3.11 The `ax.set()` command sets the axis limits manually. The axis limits in this plot were set using `ax.set(xlim=(−1, 2), ylim=(−1, 2))`.

PRACTICE PROBLEMS

3.3 Many waveforms in electrical engineering are derived from complex exponentials. Plot y = real part of $(e^{jt\pi})$ for $0 \le t \le 5$. Review Section 2.6 if you have forgotten how to find the real part of a complex number. Title the plot "Complex Exponential" and label the horizontal axis "Time, t (s)" and the vertical axis "Voltage, V (volts)". ©®

TECH TIP: OSCILLOSCOPES

Electrical engineers often use oscilloscopes to measure and graph time-varying voltages in circuits. As Fig. 3.12 illustrates, oscilloscopes are usually marked with a grid background. The user sets the scope so that each vertical division of the grid represents a certain number of volts, abbreviated V/div, and so that each horizontal division corresponds to a known amount of time, abbreviated time/div.

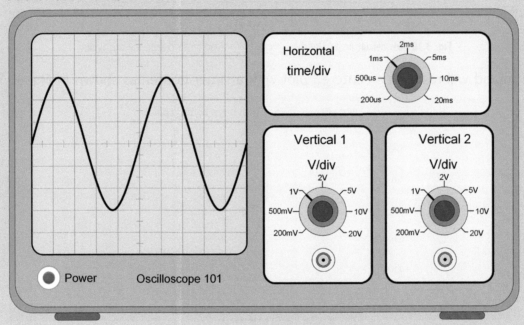

Fig. 3.12 Oscilloscope front panel. The horizontal and vertical scales are set in time/div and V/div, respectively.

3.2.8 Grids

You can add an oscilloscope-like grid to your plots with the grid method. After creating the plot, turn the grid on or off using the following commands.

```
ax.grid(visible=True)
ax.grid(visible=False)
```

For example, the plot of Fig. 3.13 is created with the following commands:

```
import numpy as np
import matplotlib.pyplot as plt
```

```
%matplotlib inline
x = np.linspace(0,np.pi,100)
y = 1 - np.power(np.sin(x),np.sin(x))

fig, ax = plt.subplots()
ax.plot(x,y)
ax.grid(visible=True)
```

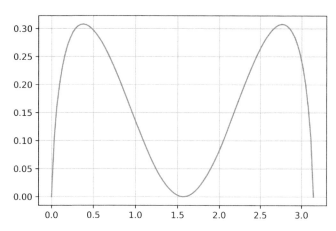

Fig. 3.13 Gridlines, when enabled with the `ax.grid(visible=True)` command, default to light gray lines at the same locations as the axis tick marks.

3.4 Create an oscilloscope-like plot, that is, one using a grid, of the waveform $x(t) = \cos(3t) - 2$ for $0 \leq t \leq 5$. Label the plot "Oscilloscope Trace", the horizontal axis "Time (s)", and the vertical axis "Volts (V)". ©®

3.2.9 Logarithmic Axis Scaling

Logarithmic axis scaling is useful in plotting values where the independent (horizontal) or dependent (vertical) variables have dense information for small values, but change progressively less for larger values of the independent (horizontal) variable. For example, attempting to plot each of the 145 standard value 5% resistor values from 1 Ω to 1 MΩ would not be readable on a standard, linearly scaled axis because the 0.1 Ω difference between the 1.0 Ω and 1.1 Ω would not be visible on a scale that extended to 1,000,000 Ω.

Chapter 3: Plotting in Python

To address this, the horizontal, vertical, or both axes of a plot may be logarithmically scaled, using the commands

```
ax.set_xscale('log')
ax.set_yscale('log')
```

Note that a logarithmically scaled axis cannot contain a zero value, since $\log(0) = -\infty$.

Plots comparing the same data with and without logarithmic scaling of the horizontal axis are shown in Fig. 3.14. Notice how horizontal log scaling of this data set reveals important information lost in the linearly scaled plot, even though the horizontal and vertical ranges of the axes are the same in both.

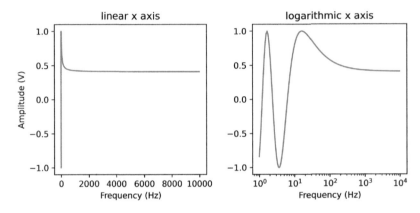

Fig. 3.14 Linear and logarithmic axis scales. For some functions and data sets, more detail is revealed using logarithmic *(right panel)* instead of linear *(left panel)* scaling. Note how using logarithmic scaling on the horizontal axis changes the tick mark locations and styling.

TECH TIP: BODE PLOTS

Bode plots are a type of plot used by electrical engineers to represent the frequency-specific amount of signal passed by a filter. Their vertical scale is measured in decibels, abbreviated as dB, which are calculated as $20 \log_{10}(x)$, where x is the amount of signal. The horizontal scale is always plotted logarithmically. For example, to generate the Bode plot shown in Fig. 3.15 from the four pairs of frequencies (Hz) and voltage (V) pairs {1, 1}, {10, 1}, {100, 0.1}, and {1000, 0.01}, use the program

```
import numpy as np
import matplotlib.pyplot as plt
```

```
%matplotlib inline
f = np.array([1, 10, 100, 1000])
y = np.array([1, 1, 0.1, 0.01]) # y axis data is given
dB = 20*np.log10(y) # convert y axis units to dB

fig, ax = plt.subplots()
ax.plot(f,dB)
ax.set_xlabel('Frequency, f (Hz)') # x label, symbol (units)
ax.set_ylabel('Magnitude, |H(f)| (dB)') # y label, symbol (units)
ax.set_xscale('log') # Scale horizontal axis logarithmically
plt.show() # Needed with Python interpreter
```

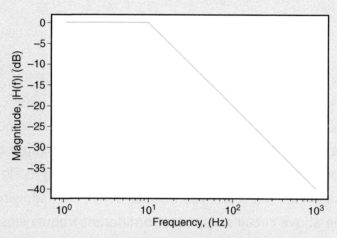

Fig. 3.15 Bode plot generated using a Python program. The horizontal axis has been set to a log scale using `ax.set_xscale('log')`.

3.5 Plot the filter response

$$y = \left(\frac{1}{10000 + f^2}\right) \tag{3.2}$$

using the styling of the Tech Tip *Bode Plots*. Use `np.logspace()` to generate logarithmically spaced points for the *f* vector array from 1 Hz to 10,000 Hz. Note that the *Bode Plots* Tech Tip provided the *f* and *y* vector arrays, but in this problem, calculate them using `np.logspace` and Eq. 3.2. Hints:

Chapter 3: Plotting in Python   113

- If `np.logspace` returns infinite numbers, reread its section in Chapter 2.4.
- Recall that `*`, `/`, and `**` work the same for NumPy arrays as they do for single numbers.
- This problem requires you to *calculate* your own horizontal and vertical vector arrays, instead of their having been given to you in the *Bode Plots* Tech Tip. The dB computation and the logarithmically scaled axis plotting routine, however, are the same as in the *Bode Plots* Tech Tip example, and indeed, for any Bode plot.

TECH TIP: LOW-PASS FILTERS

Low-pass filters are a common type of electrical circuit that removes high frequencies and allows lower ones to pass through. The simplest type can be constructed from a single resistor and capacitor, as shown in Fig. 3.16. Circuit analysis courses will teach you how to derive the following equation that models the above circuit's response to different frequencies:

$$H(f) = \frac{1}{1 + j2\pi fRC} \tag{3.3}$$

Fig. 3.16 Simple low-pass filter.

Equation 3.3 is the circuit's *transfer function H(f)* and evaluates to a complex number, as described in Chapter 2.6. To see what the circuit does to an input signal of frequency f, evaluate $H(f)$ at that frequency and take the magnitude of the resulting complex number. If $|H(f)| = 1$, it passes that frequency without change. If $|H(f)| > 1$, it amplifies the signal, and if $|H(f)| < 1$, it attenuates the signal.

Since $H(f)$ is complex, it is the magnitude $|H(f)|$ that must be plotted against frequency f. When the amplitude of this is plotted in decibels, and the frequency is plotted on a log scale, this is called a Bode plot. It is usually plotted using the conventions discussed in the previous Tech Tip: *Bode Plots*. The frequency, f, is plotted along a logarithmically scaled horizontal axis, and the magnitude, $|H(f)|$, is first converted to dB.

For example, to plot the response of this circuit with $R = 1$ kΩ and $C = 1.6$ μF for $1 \leq f \leq 10$kHz, shown in Fig. 3.17, use the following program. Note the new command `ax.grid(which='both')`, which adds gridlines to both the major and minor tick marks.

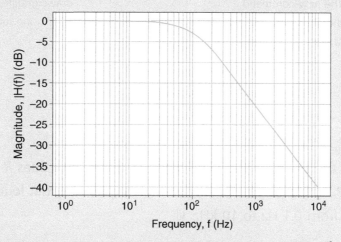

Fig. 3.17 Low-pass filter Bode plot. The low-pass filter begins to cut frequencies off roughly above $10^2 = 100$ Hz. For every 20 dB attenuation, the amplitude decreases by a factor of 0.1. Hence, by $10^3 = 1000$ Hz, only $0.1 = 10\%$ of the signal passes, and by 10 kHz, only $0.01 = 1\%$.

```
import numpy as np
import matplotlib.pyplot as plt
%matplotlib inline

R = 1000 # Resistor, R = 1 kOhm
C = 1.6e-6 # Capacitor, C = 1.6 uF

# Use 100 log spaced frequencies, f, from 1 to 10000 Hz
f = np.logspace(0,4,100) # f limits are 10**0 and 10**4

H = 1/(1+1j*2*np.pi*f*R*C) # Calculate H vector
dB = 20*np.log10(np.absolute(H)) # Calculate H magnitude in dB
```

```
fig, ax = plt.subplots()
ax.plot(f,dB)
ax.set_xlabel('Frequency, f (Hz)') # x label, symbol (units)
ax.set_ylabel('Magnitude, |H(f)| (dB)') # y label, symbol (units)
ax.set_xscale('log')
ax.grid(which='both') # Display grid - new command!
plt.show() # Needed with Python interpreter shell
```

NOTE

The new command `ax.grid(which='both')` draws gridlines at both the major and the minor ticks. This shows the effects of using log axis scaling.

PRACTICE PROBLEMS

3.6 Redo the previous Tech Tip example using the values $R = 5k\Omega$ and $C = 3.2 \, \mu F$. Plot $|H(f)|$ in dB for 100 frequencies that logarithmically span from 0.1 Hz to 10 kHz. ®

3.3 MULTIPLE-LINE PLOTS

3.3.1 Two-Line Plots

Just as oscilloscopes can display two channels of input data, engineers can graph more than one data set on the same axes, as shown in Fig. 3.18. The two data sets `{t1, y1}` and `{t2, y2}`, for example, can be plotted on the same axes, ax, using the command

```
ax.plot(t1, y1, t2, y2)
```

Consider the functions `y1 = sin(t)` and `y2 = cos(t)` defined over the closed time interval $0 \leq t \leq 6$. The program listing of Fig. 3.18 generate a time vector array `t` with 1000 linearly spaced elements from 0.0 to 6.0. Recall that both endpoints, 0.0 and 6.0, are included in the array

```
Python Interpreter                                                    >>>
>>> # Two-line plots sin(t) and cos(t)
>>>
>>> import numpy as np
>>> import matplotlib.pyplot as plt
>>>
>>> t = np.linspace(0.0, 6.0, 1000)
>>> y1 = np.sin(t)
>>> y2 = np.cos(t)
>>>
>>> fig, ax = plt.subplots()
>>> ax.plot(t, y1, t, y2)
>>>
>>> ax.set_title('Multiple Plots in a Single Axis')
>>> ax.set_xlabel('Time (s)')
>>> ax.set_ylabel('Amplitude (V)')
>>> plt.show()
```

```
Jupyter Notebook                                                    .ipynb
 1  # Two-line plots sin(t) and cos(t)
 2
 3  import numpy as np
 4  import matplotlib.pyplot as plt
 5  %matplotlib inline
 6
 7  t = np.linspace(0.0, 6.0, 1000)
 8  y1 = np.sin(t)
 9  y2 = np.cos(t)
10
11  fig, ax = plt.subplots()
12  ax.plot(t, y1, t, y2)
13  ax.set_title('Multiple Plots in a Single Axis')
14  ax.set_xlabel('Time (s)')
15  ax.set_ylabel('Amplitude (V)')
```

Fig. 3.18 Two-line plots, `sin(t)` and `cos(t)`. Each pair of arrays, `{t, y1}` and `{t, y2}`, in the axes method `ax.plot(t, y1, t, y2)` produces a line plot.

when the NumPy `linspace()` function is used. Arrays of 1000 elements are then generated for each of the functions `y1 = sin(t)` and `y2 = cos(t)` using the NumPy `sin()` and `cos()` functions. In this example, the same time vector, a one-dimensional array `t`, was used for both arrays, `y1` and `y2`, but in general they could have been different. Separate time arrays `t1` and `t2`, for example, could have been defined for each function `y1 = sin(t1)` and `y2 = cos(t2)`, respectively. The Matplotlib pyplot `subplot()` method generates a figure, called `fig`, with a set of axes called `ax` for plotting. Each pair of arrays, `{t, y1}` and `{t, y2}`, are then displayed in the figure `fig` as a separate line plot on the axes `ax` using the `ax.plot(t, y1, t, y2)` method. The resulting plot is shown in Fig. 3.19.

3.3.2 Line Options
Note that all the options introduced previously work as expected when multiple lines are plotted. For example, the command

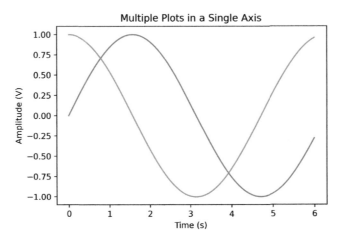

Fig. 3.19 Two-line plot of $y1 = \mathtt{sin(t)}$ and $y2 = \mathtt{cos(t)}$ over the time interval $0 \leq t \leq 6$. Two plots on the same graph can be made using the axes method, $\mathtt{plot(t1,\ y1,\ t2,\ y2)}$.

```
ax.plot(t,y1,'g',t,y2,'k--',linewidth=2)
```

will make the first line green and the second black and dashed, with each plotted at twice the default line width. This command, implemented in the listing given in Fig. 3.20, produces the plot shown in Fig. 3.21.

```
Python Interpreter                                                    >>>
>>> # Plotting sin(t) and cos(t) on the same axis
>>> # using different line colors and types.
>>>
>>> import numpy as np
>>> import matplotlib.pyplot as plt
>>>
>>> t = np.linspace(0.0, 6.0, 1000)
>>> y1 = np.sin(t)
>>> y2 = np.cos(t)
>>>
>>> fig, ax = plt.subplots()
>>> ax.plot(t,y1,'g',t,y2,'k--',linewidth=2)
>>> plt.show()
```

```
Jupyter Notebook                                                   .ipynb
 1  # Plotting sin(t) and cos(t) on the same axis
 2  # using different line colors and types.
 3
 4  import numpy as np
 5  import matplotlib.pyplot as plt
 6  %matplotlib inline
 7
 8  t = np.linspace(0.0, 6.0, 1000)
 9  y1 = np.sin(t)
10  y2 = np.cos(t)
11
12  fig, ax = plt.subplots()
13  ax.plot(t,y1,'g',t,y2,'k--',linewidth=2)
```

Fig. 3.20 Two plots with different line colors and line types having the same thickness. Multiple plot options for each plot can be included.

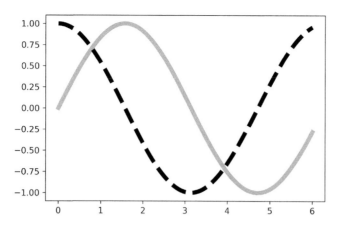

Fig. 3.21 Using different colors and line types to plot `sin(t)` and `cos(t)` on the same `axes`. Both lines have the same thickness. The command `ax.plot(t,y1,'g',t,y2,'k-',linewidth=2)` generates the expected plot lines.

3.3.3 More Than Two-Line Plots

More lines can be plotted on one set of axes, as illustrated in Fig. 3.22. For example, with more calculated values of *y3*, *y4*, and so on, one could use a command such as

```
ax.plot(t1,y1,t2,y2,t3,y3,t4,y4)
```

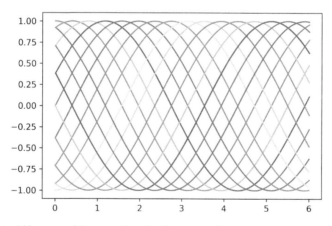

Fig. 3.22 Generating multiple plots. Adding more data array pairs, using the command `ax.plot(t1,y1,t2,y2,t3,y3,t4,y4)`, results in additional plots.

3.3.4 Adding Legends

A figure legend enables a viewer to tell the difference between different lines plotted on the same set of axes. To add a legend to a set of two lines, use the syntax

```
ax.legend(['name1', 'name2'])
```

To add a legend to the sin/cos plot in our earlier example, type

```
ax.legend(['sin(t)', 'cos(t)'])
```

To add a legend to a set of two lines, use the syntax

```
ax.legend(['Plot 1','Plot 2'], loc=code)
```

where each string is a legend label. Do not forget to enclose the list of legend strings in square brackets. The location code determines where the legend is placed. The location of the legend, by default, goes to the best location. If you wish to specify the location, use the loc = location code as indicated in Table 3.1. Legends are especially useful for printing on black and white printers. In that case, denote the difference between plots by using solid and dashed lines. The Python code listed in Fig. 3.23 demonstrates the use of legends with plots of a sin and cosine wave on the same axis, as shown in Fig. 3.24.

TABLE 3.1   Legend location codes

Location	Loc Code
Best	0
Upper right	1
Upper left	2
Lower left	3
Lower right	4
Right	5
Center left	6
Center right	7
Lower center	8
Upper center	9
Center	10

```
Python Interpreter                                          >>>
>>> # Plotting sin(t) and cos(t) on the same axis
>>>
>>> import numpy as np
>>> import matplotlib.pyplot as plt
>>>
>>> t = np.linspace(0.0, 6.0, 1000)
>>> y1 = np.sin(t)
>>> y2 = np.cos(t)
>>>
>>> fig, ax = plt.subplots()
>>> ax.plot(t,y1,t,y2)
>>> ax.set_title('Adding Legends')
>>> ax.set_xlabel('Time (s)')
>>> ax.set_ylabel('Amplitude (V)')
>>>
>>> ax.legend(['sin(t)','cos(t)'], loc = 0)
>>> plt.show()
```

```
Jupyter Notebook                                          .ipynb
 1  # Plotting sin(t) and cos(t) on the same axis
 2
 3  import numpy as np
 4  import matplotlib.pyplot as plt
 5  %matplotlib inline
 6
 7  t = np.linspace(0.0, 6.0, 1000)
 8  y1 = np.sin(t)
 9  y2 = np.cos(t)
10
11  fig, ax = plt.subplots()
12  ax.plot(t,y1,t,y2)
13  ax.set_title('Adding Legends')
14  ax.set_xlabel('Time (s)')
15  ax.set_ylabel('Amplitude (V)')
16
17  ax.legend(['sin(t)', 'cos(t)'], loc = 0)
```

Fig. 3.23 Including a legend with two plots drawn on a single axis.

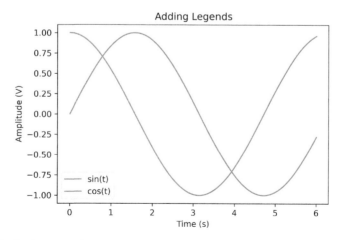

Fig. 3.24 Adding legends to multiple plots. Using the command `ax.legend(['sin(t)', 'cos(t)', loc=0)` allows legends to added for the sin(*t*) and cos(*t*) plot. The location code 0 used in this example chooses the best location.

3.7 A circuit is found to have a measured voltage described by the function $v_1(t)=2e^{-t}$ over the time interval $0 \le t \le 1$. Adding a capacitor introduces oscillations, making the new voltage response $v_2(t)=2e^{-t}\cos(20t)$ over the same time interval. Plot both $v_1(t)$ and $v_2(t)$ on the same set of axes, label the axes, and create a legend. Use as many points as needed for a smooth plot. ®

3.4 SCATTERPLOTS

Scatterplots use similar data to line plots, but instead of joining each pair of data {x, y} with lines, scatterplots mark the points themselves with markers. They are most appropriate in graphing discrete data values, where the linear interpolation between data points that are implied by connecting them with a line is not appropriate. To create scatterplots, use the same syntax as for a plot, but use one of the symbols in Table 3.2 instead of a linestyle symbol.

TABLE 3.2 Scatterplot symbol abbreviations

Symbol	Abbreviation
•	'.'
o	'o'
×	'x'
+	'+'
*	'*'
△	'^'
□	's'

To create a scatterplot for measured resistances `r = [57, 65, 63, 59, 65]` of five different heating elements `x = [1, 2, 3, 4, 5]`, first create the vectors as `NumPy arrays` and use the statement

```
ax.plot(x,r,'o').
```

to produce the scatterplot. The complete code listing is given in Fig. 3.25, and the scatterplot is shown in Fig. 3.26.

```
Python Interpreter                                          >>>
>>> # Scatter plot
>>>
>>> import numpy as np
>>> import matplotlib.pyplot as plt
>>>
>>> x = np.arange(1, 6)
>>> r = np.array([57,65,63,59,65])
>>>
>>> fig, ax = plt.subplots()
>>> ax.plot(x,r,'o')
>>>
>>> ax.set_title('Scatter Plot')
>>> ax.set_xlabel('Observation number')
>>> ax.set_ylabel('Temperature (F)')
>>>
>>> plt.show()
```

```
Jupyter Notebook                                          .ipynb
 1  # Scatter plot
 2
 3  import numpy as np
 4  import matplotlib.pyplot as plt
 5  %matplotlib inline
 6
 7  x = np.arange(1, 6)
 8  r = np.array([57, 65, 63, 59, 65])
 9
10  fig, ax = plt.subplots()
11  ax.plot(x, r, 'o')
12
13  ax.set_title('Scatter Plot')
14  ax.set_xlabel('Observation Number')
15  ax.set_ylabel('Temperature (F)')
```

Fig. 3.25 Scatterplots. Scatterplots are like line plots, but symbol characters are used to define markers instead of lines.

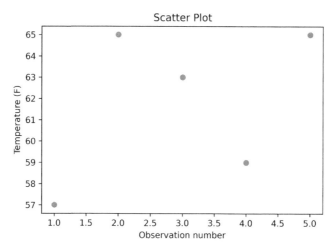

Fig. 3.26 Scatterplot of two data arrays produced by the code in Fig. 3.25. A scatterplot of the two arrays, x and r, was produced using the command `ax.plot(x,r,'o')`.

The same techniques learned earlier with `ax.plot()` to specify marker color, title, axis labeling, and grid also work for scatterplots. Now `color` specifies the marker color rather than the line color. You can also specify the marker size using `'markersize'`. For example,

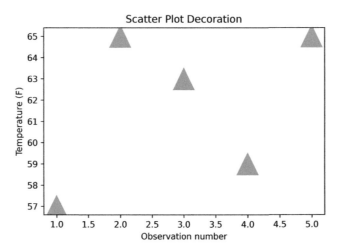

Fig. 3.27 Scatterplot with custom color and marker size.

the plot shown in Fig. 3.26 could be changed to that shown in Fig. 3.27 using the command

```
ax.plot(x, r, '^', color=[1.0, 0.5, 0.0], markersize=25)
```

PRACTICE PROBLEMS

3.8 Five 10 kW resistors are pulled from a drawer and measured. Their values are found to be 9.6, 10.2, 10.4, 9.9, and 10.0 kΩ. Scatterplot these values using circles as markers. For the horizontal axis, use the measurement numbers 1, 2, 3, 4, and 5. Label the axes and title the plot. ®

PRO TIP: PROFESSIONAL LICENSURE (PE)

Just as doctors and lawyers are licensed to practice medicine and law, engineers can be licensed as well. Licensed engineers are called "professional engineers", and they add the letters P.E after their name, such as Ken Shabby, P.E. Unlike law and medicine, most electrical engineering careers do not require licensure. However, such requirements are becoming increasingly more common. Careers in which licensure is especially helpful include the following:

- Architecture and building: Only P.E.s may certify certain types of engineering plans.
- Consulting: Only P.E.s may engage in private engineering consulting, and engineering consultant firms are required to maintain a certain minimum ratio of licensed engineers on their staff.
- Government: Being a PE. Is a common requirement for higher-level state and federal engineering jobs.
- Energy: Most power distribution companies seek P.E.s to help them certify power station plans.
- Teaching: Some states require engineering departments to ensure that a certain percentage of their engineering faculty are licensed.

Licensing requirements vary by state, but a typical path includes graduating from a four-year engineering program, passing the Fundamentals of Engineering (F.E.) exam, working for four years under a P.E., and then passing the P.E. exam.

3.5 SCRIPTS

Students often find themselves retyping many lines of previously entered commands when modifying complicated plots. Saving commands in a Jupyter Notebook code cell or text file is far more efficient.

Jupyter Notebooks have the advantage of allowing embedded markup text to document details about the figures being saved. However, one of the text editors that ship with the operating system, such as Notepad (Windows), Nano or TextEdit (MacOS), or VI (Linux), will allow one to copy and paste commands between it and the Python interpreter. Do not use Word or another word processor that saves formatting commands; they do not save pure ASCII code by default.

For example, the code listing to generate the *sinc* function (commonly found in more advanced electrical engineering courses) and a straight horizontal line shown in Fig. 3.28 is given in Fig. 3.29. Say a reviewer asks you to make the lines twice as thick. With the code below, this can be done in seconds by changing the plot line as shown below and rerunning the script or code cell.

```
ax.plot(t, y1, 'b-', t, y2, 'k-', linewidth = 5)
```

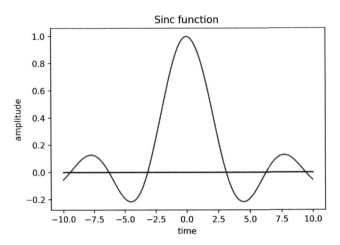

Fig. 3.28 The sinc waveform, commonly found in advanced engineering courses, and a horizontal zero line. Saving your figure-producing code in a file when creating figures allows rapid future modifications, perhaps to accommodate new data or incorporate reviewer suggestions.

```
Python Interpreter                                                    >>>
>>> # Python interpreter command listing
>>>
>>> import numpy as np
>>> import matplotlib.pyplot as plt
>>>
>>> t = np.linspace(-10, 10, 100)
>>> y1 = np.sin(t)/t
>>> y2 = np.zeros(len(y1))
>>>
>>> fig, ax = plt.subplots()
>>> ax.plot(t, y1, 'b-', t, y2, 'k-')
>>> ax.set_title('Sinc function')
>>> ax.set_xlabel('time')
>>> ax.set_ylabel('amplitude')
>>> plt.show()
```

```
Jupyter Notebook                                                  .ipynb
 1  # Jupyter notebook code cell script
 2
 3  import numpy as np
 4  import matplotlib.pyplot as plt
 5  %matplotlib inline
 6
 7  t = np.linspace(-10, 10, 100)
 8  y1 = np.sin(t)/t
 9  y2 = np.zeros(len(y1))
10
11  fig, ax = plt.subplots()
12  ax.plot(t, y1, 'b-', t, y2, 'k-')
13  ax.set_title('Sinc function')
14  ax.set_xlabel('time')
15  ax.set_ylabel('amplitude')
```

Fig. 3.29 Code listing for plotting the sinc and straight-line function. Saving your plotting commands in a file or code cell allows you to make future changes and replot your graph. Plotting commands typed to the Python interpreter need to be recalled or re-entered with changes to plot a new graph.

3.9 Create a script to build the sinc waveform without the horizontal zero line given in the previous example, and then modify the script to make the line five times as thick as the default. Run the script and show both the script and the result. ©®

3.6 LAYERING PLOTS

One of the most powerful features of `ax.plot()` is its ability to layer multiple plots on the same axes. There are two different ways to accomplish this:

1. Use the `ax.plot(t1,y1,t2,y2)` syntax introduced earlier.
2. Use separate `ax.plot()` statements, that is, `ax.plot(t1,y1)` and `ax.plot(t2,y2)`.

Let us begin by describing the weakness of the first method. You know how to plot two sets of NumPy array data, [t1, y1] and [t2, y2], with two lines using

```
ax.plot(t1, y1, t2, y2)
```

You also know how to format them individually — for instance, making the first line red and solid and the second line black and dotted. Recall that the commands must be grouped with the data. For example,

$$\underset{\text{first line}}{\underbrace{\texttt{ax.plot(t1,y1,'r-'}}},\underset{\text{second line}}{\underbrace{\texttt{t2,y2,.,'k :')}}}$$

Consider plotting the functions

$$y_1(t) = \cos(2\pi t)e^{-t/5} \tag{3.4}$$

and

$$y_2(t) = e^{-t/5} \tag{3.5}$$

over the same time interval, $0 \leq t \leq 15$. These two functions can be generated using the code statements

```
import numpy as np
import matplotlib.pyplot as plt
# %matplotlib inline needed with a Jupyter Notebook

t = np.linspace(0,15,1000)
y1 = np.cos(2*np.pi*t) * np.exp(-t/5)
y2 = np.exp(-t/5)

fig, ax = plt.subplots()
ax.plot(t,y1,'r-',t,y2,'k:')
plt.show() # Include this with the Python interpreter
```

to produce the plot shown in Fig. 3.30.

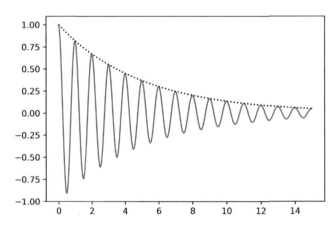

Fig. 3.30 Formatting two plots on the same graph with different line colors and line types. Note that both lines have the same thicknesses.

In Fig. 3.30 the black dotted line is hard to see because it is small. How can you make the first, red line thick and the second, black line thin? Two-part commands such as `linewidth=5` and `color=[1, 0, 1]` will not work because they must appear at the very end of the command and apply to all line plots when the `ax.plot()` command is used. For example, the plot statement

$$\texttt{ax.plot(}\underbrace{\texttt{t, y1,' r-'}}_{\text{first line}}\texttt{,}\underbrace{\texttt{t, y2,' k :'}}_{\text{Second line}}\texttt{,}\underbrace{\texttt{linewidth = 5}}_{\text{both lines}}$$

will produce the plot shown in Fig. 3.31.

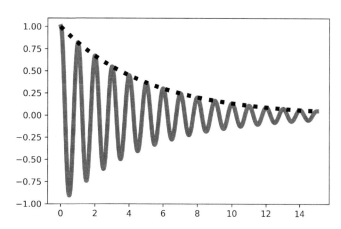

Fig. 3.31 The statement `ax.plot(t,y1,'r-',t,y2,'k:',linewidth=5)` makes both lines the same thickness.

To obtain the desired effect of just the second black line being thicker, one must use two separate `plot()` commands to layer one line over the other — in this case, the two statements

```
ax.plot(t,y1,'r-',linewidth=1)
ax.plot(t,y2,'k:',linewidth=7)
```

will produce the desired result shown in Fig. 3.32.

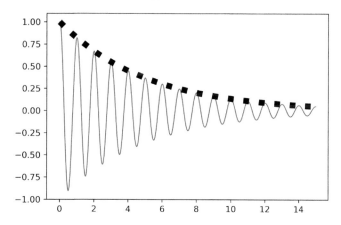

Fig. 3.32 Using two plot statements to layer the plots with independent line thicknesses. Layering the plot using two separate plot statements, `ax.plot(t,y1,'r-',linewidth=1)` and `ax.plot(t,y2,'k:',linewidth=7)`, will give each line its own thicknesses.

PRACTICE PROBLEMS

3.10 Create a plot of $h(t) = e^{-t}$ from $0 \leq t \leq 5$ using a black line 15 times thicker than the default. Using a separate plot command, plot the default thin line of $y(t) = e^{-t} + 0.04 \cos(15t)$ over it in white (that is, 'w'). ®

3.7 BAR PLOTS

A bar plot is a convenient way to visualize large quantities of raw data grouped into bins. For example, consider a large class that completes a lab and measures the capacitor voltages listed in Table 3.3.

TABLE 3.3 Capacitor voltage measurement summary

V_C	3.3	3.4	3.5	3.6	3.7
Number of observations, n	3	12	34	25	9

To graph this relationship, use the same syntax as the plot command, but use `ax.bar(x,y,width)` (where the width argument is optional). As always, the horizontal data goes first, followed by the vertical data. The last argument, `width`, in `ax.bar` specifies how wide each bar should be. The code listing in Fig. 3.33 produces the bar plot shown in Fig. 3.34, which summarizes the data in Table 3.3.

```
Python Interpreter                                          >>>
>>> # Capacitor voltage measurement bar plot
>>>
>>> import numpy as np
>>> import matplotlib.pyplot as plt
>>>
>>> Vc = np.arange(3.3, 3.8, 0.1) # [3.3, 3.4, 3.5, 3.6, 3.7]
>>> n = np.array([3, 12, 34, 25, 9])
>>>
>>> fig, ax = plt.subplots()
>>> ax.bar(Vc, n, 0.08)
>>>
>>> ax.set_title('Capacitor Voltages')
>>> ax.set_xlabel('Voltage')
>>> ax.set_ylabel('Number')
>>>
>>> plt.show()
```

```
Jupyter Notebook                                          .ipynb
 1  # Capacitor voltage measurement bar plot
 2
 3  import numpy as np
 4  import matplotlib.pyplot as plt
 5  %matplotlib inline
 6
 7  Vc = np.arange(3.3, 3.8, 0.1) # [3.3, 3.4, 3.5, 3.6, 3.7]
 8  n = np.array([3, 12, 34, 25, 9])
 9
10  fig, ax = plt.subplots()
11  ax.bar(Vc, n, 0.08)
12
13  ax.set_title('Capacitor Voltages')
14  ax.set_xlabel('Voltage')
15  ax.set_ylabel('Number');
```

Fig. 3.33 Capacitor voltage measurement `bar plot`. Bar plots are produced using the `ax.bar(x,y,width)` command, where `x` is the horizontal data, `y` the vertical data, and the optional `width` specifies the bar thickness.

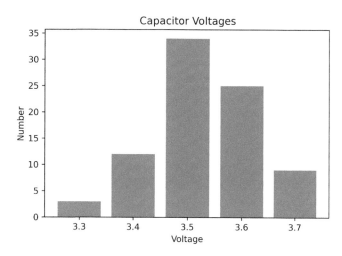

Fig. 3.34 Single `bar` plot summarizing capacitor voltage measurements. The horizontal axis represents the capacitor voltage and the vertical axis the number of measurements recorded with that voltage.

The same techniques used to adorn line plots also work with `bar` plots. For example, the `bar` plot presented in Fig. 3.33 has `title` and `axis labels` added using `ax.set_title()`, `ax.set_xlabel()`, and `ax.set_ylabel()`. In addition, the command `ax.set(xlim = (xmin, xmax), ylim = (ymin, ymax))` could be used to change the `axis limits` to add more white space to the sides of the plot.

Multiple `bar` plots can be colored grouped into different series. Consider the two sets of observations summarized in Table 3.4. To present both sets of `bar` graphs on the same plot, set the `bar width` to 2.5 volts. Then shift the test 1 `bar` graph to the left by 1.5 volts and the test 2 `bar` graph to the right by 1.5 volts. If the test 1 `bar` graph is colored blue and the test 2 `bar` graph colored red, it will be easy to distinguish them. The code listing to generate this multiple `bar` plot is given in Fig. 3.35, and the resulting plot is presented Fig. 3.36.

TABLE 3.4 Voltage measurements from two different sets of observations

V	60	70	80	90	100
Number of test 1 observations	3	12	34	25	9
Number of test 2 observations	4	10	30	24	7

```
Python Interpreter                                              >>>
>>> # Two observation data set bar plot
>>>
>>> import numpy as np
>>> import matplotlib.pyplot as plt
>>>
>>> V = np.array([60, 70, 80, 90, 100])
>>> n1 = np.array([3, 12, 34, 25, 9])
>>> n2 = np.array([4, 10, 30, 24, 7])
>>>
>>> fig, ax = plt.subplots()
>>> ax.bar(V-1.5, n1, color='b', edgecolor='k', width=2.5)
>>> ax.bar(V+1.5, n2, color='r', edgecolor='k', width=2.5)
>>> ax.legend(['test 1', 'test 2'])
>>> plt.show()
```

```
Jupyter Notebook                                              .ipynb
 1 | # Two observation data set bar plot
 2 |
 3 | import numpy as np
 4 | import matplotlib.pyplot as plt
 5 | %matplotlib inline
 6 |
 7 | V = np.array([60, 70, 80, 90, 100])
 8 | n1 = np.array([3, 12, 35, 25, 10])
 9 | n2 = np.array([4, 10, 30, 24, 7])
10 |
11 | fig, ax = plt.subplots()
12 | ax.bar(V-1.5, n1, color='b', edgecolor='k', width=2.5)
13 | ax.bar(V+1.5, n2, color='r', edgecolor='k', width=2.5)
14 | ax.legend(['test 1', 'test 2'])
```

Fig. 3.35 A bar plot with two data sets. Similarly to line plots, two or more sets of bar plots can be presented on the same figure by using more than one ax.bar(x,y,d) command. Notice how the two bar plots are shifted, colored, and their bar width set to make them more visually appealing.

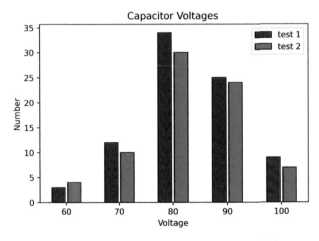

Fig. 3.36 A bar plot with two data sets. To plot more than one set of bar plot data, use multiple ax.bar(x,y,width) commands. Shifting each bar plot, using different colors, and adding a legend can improve the data presentation.

3.11 Digital logic gates are the building blocks of digital computers, and their function will be introduced in future courses. These components have inputs and outputs with a (usually undesired) delay between when an input signal changes and when the output responds, called the propagation delay t_p. We will discuss these gates in greater detail in Section 5.9; however, Table 3.5 summarizes the propagation delays found for one common type of gate: the NOT gate on the 74LS04 chip.

Create a bar plot displaying the values in Table 3.5. Label the axes and give it a title. Hint: Rather than converting a very small number such as 2.5 ns to 0.0000000025 s, leave it as 2.5 and label the appropriate axis as "time (ns)". ®

TABLE 3.5 Digital logic propagation delay summary for the 74LS04 NOT gate

t_p	2.5 ns	2.75 ns	3 ns	3.25 ns	3.5 ns
# of chips	2	12	27	16	3

3.8 SUBPLOTS

You have learned how to include multiple plots on the same set of axes using one or more axes methods such as `plot()` or `bar()`. Sometimes you may want to group multiple `axes subplots` in the same figure. Fig. 3.37, for example, shows the frequency response of a Butterworth low-pass filter plotted against various combinations of x and y `axis scales`. Note that each `axis` has a different scale, so it is not possible to overlay all plots on the same axes.

To produce multiple axes subplots, first specify the number of rows and columns of axes using the `subplots(r,c)` command. You have used this command previously to return figure and axes objects when you entered `fig, ax = plt.subplots()`. When subplot is given arguments of

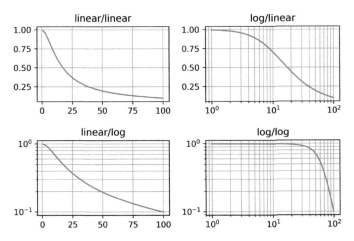

Fig. 3.37 Multiple axes subplots on the same figure. More than one axes subplot is needed because each graph has a different set of linear or log scales on its x axis and y axis.

r and c, it divides the figure into r rows and c columns of axes. Fig. 3.37, for example, has two rows and two columns of axes that are generated using the command

```
fig, axs = plt.subplots(2, 2)
```

The axes are selected by indexing them as rows and columns starting with zero as illustrated in Fig. 3.38.

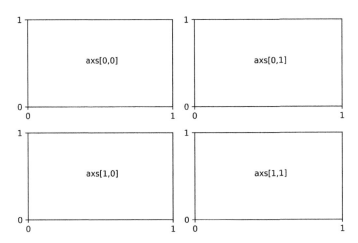

Fig. 3.38 A figure with an array of axes is created using the `subplots(rows, columns)` command. In this case, the command `fig, axs = plt.subplots(2,2)` creates a figure, fig, and axes, axs, that is a two-by-two array of axes.

As an example, using just a column array, the commands listed in Fig. 3.39 create a set of two long, narrow plots, one over the other. The top one graphs $v_1(t) = \sin(t)$ and the bottom graphs $v_2(t) =$ a triangle wave for $0 \leq t \leq 4\pi$, as shown in Fig. 3.40.

```
Python Interpreter                                              >>>
>>> # Multiple axes plot of sin and triangle waves
>>>
>>> import numpy as np
>>> import matplotlib.pyplot as plt
>>>
>>> t1 = np.linspace(0, 4*np.pi, 1000)
>>> v1 = np.sin(t1)
>>>
>>> t2 = np.linspace(0, 4*np.pi, 9)
>>> v2 = np.array([0, 1, 0, -1, 0, 1, 0, -1, 0])
>>>
>>> fig, axs = plt.subplots(2, 1)
>>>
>>> axs[0].plot(t1, v1)
>>> axs[0].set_title('sin(t)')
>>>
>>> axs[1].plot(t2, v2)
>>> axs[1].set_title('triangle wave')
>>>
>>> plt.show()
```

```
Jupyter Notebook                                              .ipynb
 1  # Multiple axes plot of sin and triangle waves
 2
 3  import numpy as np
 4  import matplotlib.pyplot as plt
 5  %matplotlib inline
 6
 7  t1 = np.linspace(0, 4*np.pi, 1000)
 8  v1 = np.sin(t1)
 9  t2 = np.linspace(0, 4*np.pi, 9)
10  v2 = np.array([0, 1, 0, -1, 0, 1, 0, -1, 0])
11
12  fig, axs = plt.subplots(2, 1)
13
14  axs[0].plot(t1, v1)
15  axs[0].set_title('sin(t)')
16
17  axs[1].plot(t2, v2)
18  axs[1].set_title('triangle wave');
```

Fig. 3.39 Multiple axes sin and triangle wave plot. The command `plt.subplots(2,1)` returns two rows and one column of axes. The axes are indexed starting at 0.

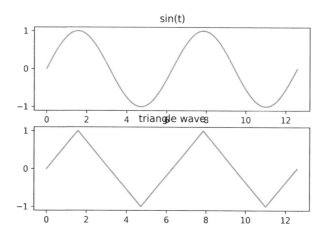

Fig. 3.40 Multiple axes sin and triangle wave figure using two rows and one column. The `plt.subplots(2,1)` command returns two sets of axes, one for the two rows and single column. Note, however, that the axes overlap each other. This can be corrected using the `plt.tight_layout()` command.

Unfortunately, the default settings for `subplots()` make individual graphs so close that their titles collide. Fix this by adding a `plt.tight_layout()` command at the end of the previous set of commands to make them space correctly, as shown in Fig. 3.41.

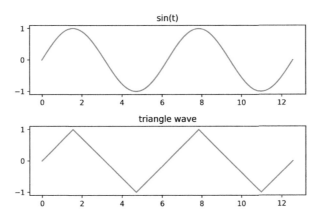

Fig. 3.41 Adding a `plt.tight_layout()` command. The `plt.tight_layout()` method spaces the individual plots so that they do not overlap. If you are using a Python interpreter, include this statement before `plt.show()`, as shown in Fig. 3.39.

PRACTICE PROBLEMS

3.12 A circuit with inductors, capacitors, and resistors can "ring" or oscillate in some cases, when driven by a voltage step. Both voltage and current oscillate with the same frequency but with different phases, such as

$$v(t) = e^{-t/4} \cos(4t) \tag{3.6}$$

and

$$i(t) = 0.01 \, e^{-t/4} \cos(4t + \pi/2) \tag{3.7}$$

Plot Eqs. 3.6 and 3.7 over the interval $0 \leq t \leq 15$ as two subplots, one on top of the other, like the example shown in Fig. 3.41. ®

3.13 Using the `ax.plot(x,y1,x,y2)` command, plot Eqs. 3.6 and 3.7 using a single axis. Which of these two methods is a better way to visualize the similarities between $v(t)$ and $i(t)$? ®

3.9 TEXT ANNOTATIONS

To create text at an arbitrary position on a plot, use the command

```
ax.text(t, y, 'text string')
```

For example, the function

$$v(t) = \frac{t}{2}\cos(t) \tag{3.8}$$

can be generated over the interval $0 \le t \le 5$ using

```
t = np.linspace(0, 5, 1000)
y = t/2 * np.cos(t)
```

The function can be plotted using

```
fig, ax = plt.subplots()
ax.plot(t,y)
```

To improve the appearance, set the axis limits of the horizontal t axis from 0 to 5 and those of the vertical y axis from -2 to 1 using the axis command

```
ax.set(xlim=(0,5), ylim=(-2,1))
```

Add text to highlight the local maxima and global minima using the commands

```
ax.text(0.4, 0.4, 'Local maximum')
ax.text(2.8,-1.8, 'Global minimum')
```

The complete graph is shown in Fig. 3.42.

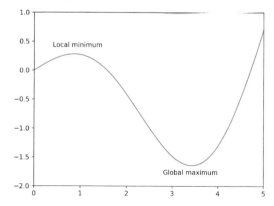

Fig. 3.42 Plot of $(t/2)\cos(t)$ with text annotations. Text annotations were added using `ax.text(t, y, 'text sting')` and the `axis limits` set using `ax.axis([0.0, 5.0, -2.0, 1.0])`.

Electrical engineers commonly use many of the 24 letters in the Greek alphabet, in both upper and lower case. It is important to be able to recognize them and pronounce their names. Table 3.6 pairs some Greek letters with their common ECE applications.

TABLE 3.6 Greek letters and their ECE applications

α	alpha	exponential decay rate
β	beta	transistor current gain
Γ	Gamma	transmission reflection
Δ	Delta	change
δ	delta	impulse function
η	eta	efficiency ratio
θ	theta	angle
λ	lambda	exponential decay rate
μ	mu (micro)	10^{-6}
ξ	xi	damping ratio
Π	Pi	product
π	pi	3.14159…
Σ	Sigma	sum
σ	sigma	conductivity, complex number real part
τ	tau	time constant
Φ	Phi	magnetic flux
φ	phi	angle (similar use of θ)
Ω	Omega	unit of resistance
ω	omega	radian frequency

3.10 ADVANCED TEXT FORMATTING

Electrical engineering plots often require exponents, subscripts, or the Greek symbols listed previously in the Tech Tip: *Greek Letters*. The Python `Matplotlib pyplot` module, therefore, provides a way for the axes methods `set_title()`, `set_xlabel()`, `set_ylabel()`, and `text()` to display them.

3.10.1 Greek Symbols

Greek symbols are very common in electrical engineering. The string is preceded with `r` to denote a raw string. The actual Greek symbol is embedded using a pair of enclosing dollar signs, $, and a backslash character, \, followed by the name of the Greek symbol. In Fig. 3.43, for example, if axes are created with the name `ax`, the commands

```
ax.set_xlabel(r'Frequency, rad/s ($\omega$)')
ax.set_ylabel(r'Resistance, ohms ($\Omega$)')
```

produce the labels "Frequency, rad/s (ω)" and "Resistance, ohms (Ω)" for the horizontal and vertical axis, respectively. Notice that the upper-case Greek letter Ω is distinguished from the lower-case Greek letter ω by the capitalization of the symbol name's first letter.

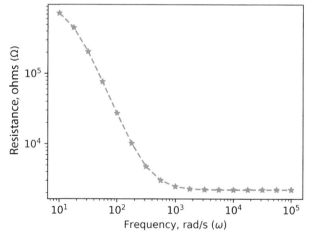

Fig. 3.43 Using Greek letters in plot labels. Use a pair of enclosing dollar signs, $, and a backslash character, \, followed by the name of the Greek symbol to make Greek letters. In this example, Ω was produced using `$\Omega$` and ω with `$\omega$`.

3.10.2 Super- and Subscripts

Superscripts, such as t^2, and subscripts, such as V_1, are easy to create. If the superscript or subscript is just a single character, use ^ or _, respectively. Figure 3.44 illustrates a single-character subscript and superscript produced using the command

```
ax.set_title(r'$V_1(t) = f(t^2)$')
```

to create the title $V_1(t) = f(t^2)$.

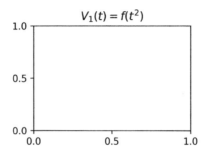

Fig. 3.44 Superscripts and subscripts in plot labels. Using the hat, $\hat{}$, and underscore, _, produce superscripts and subscripts of single characters, respectively.

To group multiple characters in the superscript or subscript, surround them with curly braces, {}. Consider the command

```
ax.text(0.45, 0.50, r'$f_{2\alpha} = e^{j\omega t}$')
```

that creates the text annotation shown in Fig. 3.45. Notice how Greek symbols can also be included in the groupings.

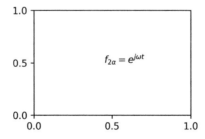

Fig. 3.45 Grouping multiple characters in a superscript or subscript. Multiple characters can be used in superscript and subscripts using curly braces, {}, to group them.

3.14 Plot the function $v_1(t) = 2e^{-\lambda t}$ over the interval $0 \le t \le 0.5$, where $\lambda = 4$. Label the vertical axis "$v_1(t)$" and give it the title "$2e^{-\lambda t}, \lambda = 4$". ®

3.11 THREE-DIMENSIONAL PLOTS

Although the two-dimensional line, scatter, and bar plots described so far are the most common types of graphics used by electrical engineers, three-dimensional plots can show how one dependent variable changes as a function of two independent variables. For example, the sinc function discussed earlier in Section 3.5 also exists as a function of two independent variables and is used in image processing. This two-dimensional sinc function is described by the equation

$$z(x,y) = \frac{\sin\left(\sqrt{x^2 + y^2}\right)}{\sqrt{x^2 + y^2}} \qquad (3.9)$$

where the dependent variable $z(x,y)$ is a function of the two independent variables x and y.

To render the function $z(x,y)$ as a two-dimensional surface in three-dimensional space, start by importing the NumPy and Matplotlib pyplot packages using the statements

```
import numpy as np
import matplotlib.pyplot as plt
```

Matrix NumPy arrays for the x and y coordinates, need to be created using meshgrid(), which can be thought of as the two dimensional counterpart of linspace(). The first argument for meshgrid() is the vector NumPy array of x locations, and its second argument is the vector NumPy array of y locations. For example,

```
x_locations = np.linspace(-15, 15, 50)
y_locations = np.linspace(-15, 15, 50)
(x, y) = meshgrid(x_locations, y_locations)
```

specifies 50 linearly spaced values ranging from −15 to 15 for the two vector `NumPy arrays` `x_locations` and `y_locations`. The `meshgrid()` method creates matrix `NumPy arrays x`, and `y`, that are used in the following surface rendering methods. If you adjust the number of points to a value other than 50, use an even number. This avoids the value `(x, y) =` `(0, 0)`, where a zero-divide error will occur for the two-dimensional sinc function as it is currently defined. We will address this issue in Chapter 5 when we discuss conditional statements.

The two-dimensional sinc function is created using

```
r = np.sqrt(x**2 + y**2)
z = np.sin(r) / r
```

The variable `r` simplifies the calculation of the sinc function using a substitution.

To generate a figure with a set of axes use

```
fig = plt.figure(figsize = (6.4, 4.8))
ax = plt.axes(projection = '3d')
```

The `figsize(width, height)` method sets the width and height of the figure. In this example, the figure width is set to 6.4 inches and the height to 4.8 inches. Although 6.4 inches by 4.8 inches is the default figure size, you can try altering these values to see how they change the plot presentation. The axes are set for a 3d projection. To plot the surface use

```
ax.plot_surface(x, y, z)
```

Labels can be added, similarly to previous plots, using

```
ax.set_xlabel('x')
ax.set_ylabel('y')
ax.set_zlabel('sinc(x,y)')
```

The complete program listing, using both the Python interpreter and a Jupyter Notebook code cell, is given in Figs. 3.46a and 3.46b and the resulting surface plot is shown in Fig. 3.47.

```
Python Interpreter                                          >>>
>>> # Two-dimensional sinc function plot
>>>
>>> # Import the required packages
>>> import numpy as np
>>> import matplotlib.pyplot as plt
>>>
>>> # Create datasets for the 2D sinc function
>>> x_locations = np.linspace(-15, 15, 50)
>>> y_locations = np.linspace(-15, 15, 50)
>>> (x, y) = np.meshgrid(x_locations, y_locations)
>>> r = np.sqrt(x**2 + y**2)
>>> z = np.sin(r) / r  # Create the 2d sinc function
>>>
>>> # Create figure and axes
>>> fig = plt.figure(figsize = (6.4, 4.8)) # Figure size
>>> ax  = plt.axes(projection = '3d')  # Create 3d projection
>>>
>>> # Plot the surface and add labels
>>> ax.plot_surface(x, y, z)
>>> ax.set_xlabel('x')
>>> ax.set_ylabel('y')
>>> ax.set_zlabel('sinc(x,y)')
>>> plt.show()
```

Fig. 3.46a Surface plot of the 2D sinc function using the Python interpreter.

```
Jupyter Notebook                                          .ipynb
 1  # Two-dimensional sinc function plot
 2
 3  # Import the required packages
 4  import numpy as np
 5  import matplotlib.pyplot as plt
 6
 7  %matplotlib inline
 8
 9  # Create datasets for the 2D sinc function
10  x_locations = np.linspace(-15, 15, 50)
11  y_locations = np.linspace(-15, 15, 50)
12  (x, y) = np.meshgrid(x_locations, y_locations)
13  r = np.sqrt(x**2 + y**2)
14  z = np.sin(r) / r   # Create the 2d sinc function
15
16  # Create figure and axes
17  fig = plt.figure(figsize=(6.4, 4.8)) # Figure size 6.4"x4.8"
18  ax  = plt.axes(projection='3d')  # Create 3d projection axes
19
20  # Plot the surface and add labels
21  ax.plot_surface(x, y, z)
22  ax.set_xlabel('x')
23  ax.set_ylabel('y')
24  ax.set_zlabel('sinc(x,y)'); # Suppress output from last statement
```

Fig. 3.46b Surface plot of the 2D sinc function using a Jupyter Notebook code cell. Small changes to this program can be used to generate color surface and wireframe plots.

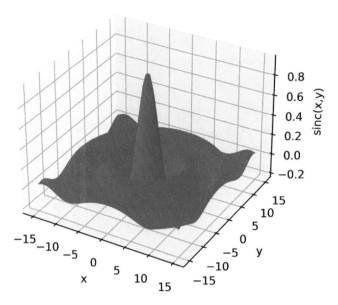

Fig. 3.47 Surface plot of the two-dimensional sinc function.

To add color to the surface plot, first import `matplotlib.cm`, and then set the `cmap` attribute to `cm.jet` in the `ax.plot_surface()` method

```
import matplotlib.cm as cm
ax.plot_surface(x, y, z, cmap=cm.jet)
```

The resulting surface, shown in Fig. 3.48, uses color to help visualize the height of the surface more clearly. Compare Fig. 3.48 with Fig. 3.47.

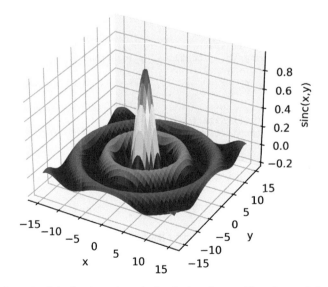

Fig. 3.48 Surface plot of the two-dimensional sinc function enhanced using the jet color map. The color map helps visualize the height of the function in many cases.

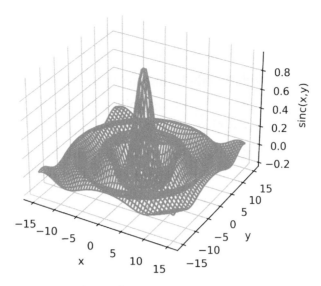

Fig. 3.49 Wireframe plot of the two-dimensional sinc function. Using the `plot_wireframe()` command in place of the `plot_surface()` command generates a wireframe plot instead of a solid surface plot.

As an alternative, the wireframe plot shown in Fig. 3.49 can be obtained by replacing the `ax.plot_surface()` method with

```
ax.plot_wireframe(x, y, z)
```

The wireframe plot gives a different visualization of the two-dimensional sinc function. Your choice of surface, colored surface, or wireframe will depend on the function and your personal preference.

PRACTICE PROBLEMS

3.15 For the following function, over the domains of both x and y from −10 to 10:

$$z(x,y) = \sin\left(\frac{x}{3}\right)\cos\left(\frac{y}{3}\right) \qquad (3.10)$$

a) create a wireframe plot ®
b) create a surface plot ®

3.12 SAVING PLOTS IN DIFFERENT FILE FORMATS

A common use of Python is to generate plots to be included in professional reports. Python graphics can be copied and pasted directly from a computer monitor and incorporated into a report using a screen-grabbing package, such as Snip-and-Sketch, bundled into Windows, or Shift-Command-4 in Macs, as done in Chapter 1. These tools, however, produce blurry images when the documents are printed, since their resolution is tied to the resolution of your monitor. The highest-quality images are produced using Python commands that save your plots to a image file of arbitary resolution.

There are two formats to consider when saving your graphics files in Python: .svg and .png. Scalable computer graphics, or .svg, files have the advantage of being saved in vector graphics format, so they can be scaled with no loss of resolution to a very compact file size. Unfortunately, fewer applications recognize this format than other formats, and many render .svg files in slightly different ways. An alternative, more universally recognized format, is the portable network graphics, or .png, file format. This format is rasterized at a predetermined number of dots per inch (dpi) when created. It will appear pixelated if rendered to a too-low resolution; but 600 dpi is generally high enough so that pixelation is not visible. A 600 dpi .png file, however, has a file size about six times greater than that of the corresponding .svg file. The command `figsize(width, height)` is needed to establish the width and the height of the image in inches so that the text is set to the correct size. By default, files will be saved to the current working directory. The following program provides an example of a simple graphic figure saved as a default resolution .png file `'pngFile.png'` and a 600 dpi .png file `'png600dpiFile.png'`. The image files are illustrated in Fig. 3.50. Notice how the 600 dpi .png compares to the default resolution .png file. To save your file in an .svg format, use a .svg instead of .png extension in the `savefig()` command.

NOTE

Using the `savefig()` method is useful if your code produces multiple plots, since it can save them all rather than requiring tediously copying and pasting each individually. Using loops (covered in Chapter 5), you can even change the filename for a sequence of test cases.

```
import numpy as np
import matplotlib.pyplot as plt
%matplotlib inline

t = np.linspace(0.0, 5.0, 1000)
y = t / 2.0 * np.cos(t)
fig, ax = plt.subplots(figsize=(3.0,2.0))
ax.plot(t,y)

ax.set_title('png default resolution')
plt.savefig('pngFile.png')

ax.set_title('png dpi=600')
plt.savefig('png600dpiFile.png', dpi=600)
```

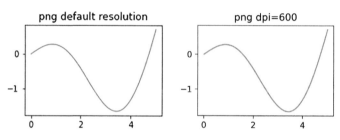

Fig. 3.50 Using `savefig()` to save plots in different file formats. Saving a figure as a file with a .png ending saves it in a portable network graphics format. When saving a .png file, you can adjust the dots per inch, or dpi, by including a `dpi=resolution` attribute. Setting the resolution to 600 dpi avoids the pixelation noise associated with the default resolution.

PRO TIP: IEEE CODE OF ETHICS

Electrical engineers must be held to a high ethical code, since their work often directly involves public safety. Safety-critical devices such as airplane transponders, hospital intravenous pump regulators, and electric vehicle braking systems are designed by engineers under multiple constraints including safety, cost, and time. The professional society of electrical engineers, the IEEE, has published a Code of Ethics to guide and protect its members. To follow the IEEE Code of Ethics, electrical engineers agree

1. To uphold the highest standards of integrity, responsible behavior, and ethical conduct.

a. to hold paramount the safety, health, and welfare of the public, to strive to comply with sustainable development practices, to protect the privacy of others, and to disclose promptly factors that might endanger the public or the environment;

b. to improve the understanding by individuals and society of the societal implications of conventional and emerging technologies, including intelligent systems;

c. to avoid real or perceived conflicts of interest whenever possible, and to disclose them to affected parties when they do exist;

d. to avoid unlawful conduct, and to reject bribery in all its forms;

e. to seek, accept, and offer honest criticism, to acknowledge and correct errors, to be honest and realistic in stating claims or estimates, and to credit the contributions of others;

f. to maintain and improve technical competence and to undertake tasks for others only if qualified by training or experience, or after full disclosure of pertinent limitations;

2. To treat all persons fairly and with respect, to not engage in harassment or discrimination, and to avoid injuring others.

a. to treat all persons fairly and with respect, and to not engage in discrimination based on characteristics such as race, religion, gender, disability, age, national origin, sexual orientation, gender identity, or gender expression;

b. to not engage in harassment of any kind, including sexual harassment or bullying behavior;

c. to avoid injuring others, their property, reputation, or employment by false or malicious actions, rumors or any other verbal or physical abuses;

3. To strive to ensure this code is upheld by colleagues and co-workers.

a. to support colleagues and co-workers in following this code of ethics, to strive to ensure the code is upheld, and to not retaliate against individuals reporting a violation.

See the unabridged text at https://www.ieee.org/about/ethics.html.

COMMAND REVIEW

Import Packages for Data and Plot Rendering

```
import numpy as np
```
Imports numpy with alias np.

```
import matplotlib.pyplot as plt
```
Imports matplotlib.pyplot with alias plt.

```
import matplotlib.cm as cm
```
Needed for 3D surface plots with colormaps.

Jupyter Notebook Plot Rendering

```
%matplotlib inline
```
Needed for Jupyter Notebook plots.

Python Interpreter Plot Rendering

```
plt.show()
```
Show the plot using the Python interpreter.

Creating Plot Data Arrays

```
x = np.array([2, 3, 7])
```
NumPy vector array x with elements 2, 3, 7.

```
np.arange(start, stop, step)
```
Linear step from start to just before stop.

```
np.linspace(start, stop, N)
```
N linearly-spaced data from start to stop.

```
np.logspace(a, b, N)
```
N log-spaced data from 10^a to 10^b.

Line Plots

```
fig, ax = plt.subplots()
```
Create figure and axes called fig and ax.

```
ax.plot(x, y)
```
Plot array y as a function of array x.

```
ax.plot(x1, y1, x2, y2)
```
Plot two lines with (x1, y1) and (x2, y2).

```
ax.plot(x1, y1, x2, y2, ...)
```
Multiple line plots of (x1, y1), (x2, y2), ...

Multiple Overlayed Plots

```
ax.plot(x,y,'r-',linewidth=5)
ax.plot(x,y+1,'g:',linewidth-2)
```
These two commands overlay plots with different color, styles, and linewidths.

```
ax.legend('text',loc=1)
```
Place legend text at location 1. See Table 3.1.

Subplots

```
fig, axs = plt.subplots(2, 2)
```
A figure, fig, with 2 × 2 subplots axes, axs.

```
axs[0,1].plot(x, y)
```
plot y as function of x on axes axs[0,1].

```
plt.tight_layout()
```
Fixes spacing when subplots overlap.

Scatterplots

```
ax.scatter(x, y)
```
Create a scatterplot of `(x, y)` data set.

```
ax.scatter(x, y, s=50,
  facecolor = 'blue', edgecolor='k')
```
Use large, blue-filled, black-edged markers.

Bar Plots

```
ax.bar(x, y)
```
Create a bar plot of the `(x, y)` data.

Options for Line, Scatter, and Bar Plots

```
ax.plot(x, y, color='blue',
  linewidth = 5, linestyle = '-')
```
Blue dashed line 5× normal width of `y` versus `x`.

Axis Options

```
ax.set_xscale('log')
```
Change x scaling from linear to `log`.

```
ax.set_yscale('log')
```
Change y scaling from linear to `log`.

```
ax.set(xlim=(xmin, xmax),
  ylim=(ymin, ymax))
```
Set axis limits.

```
ax.xaxis.set_ticks([0, 1, 2])
```
Sets `ticks` on the `x` axis to `[0, 1, 2]`.

```
ax.yaxis.set_ticks([])
```
Remove `y` axis ticks.

Text on Line, Scatter, and Bar Plots

```
ax.set_title('Plot Title')
```
Print `title` text `'Plot Title'`.

```
ax.set_xlabel('x axis text')
```
Print `'x axis text'` below horizontal `axis`.

```
ax.set_ylabel('y axis text')
```
Print `'y axis text'` left of vertical `axis`.

```
ax.text(2.0, 3.0, 'my text')
```
Print `'my text'` at location `(2.0, 3.0)`.

Special Text (Greek, Math, Superscripts, Subscripts)

```
ax.set_title(r'$\alpha$')
```
Greek characters. Note the "r" prefix.

```
ax.set_xlabel(r'$\beta_1$')
```
Subscript β_1.

```
ax.set_title(r'$e^{-t}$ + $\beta_1$')
```
Title reads: $e^{-t} + \beta_1$.

Import Statements for Three-Dimensional Plots

```
from mpl_toolkits import mplot3d
```
`import mplot3d` for 3D rendering.

```
import numpy as np
```
Needs the `NumPy` package.

```
import matplotlib.pyplot as plt
```
Needs the `matplotlib.pyplot` module.

Three-Dimensional Plot Rendering

```
ax = plt.axes(projection='3d')
```
Specify that the axes will be a 3D projection.

```
x_loc = np.linspace(-1, 1, 50)
```
Linearly space *x* axis with 50 values from −1 to 1.

```
y_loc = np.linspace(-2, 2, 50)
```
Linearly space y axis with 50 values from −2 to 2.

```
(mx, my) = np.meshgrid(x_loc, y_loc)
```
Create `arrays`, `mx` and `my`, for 3D plotting.

```
mz = np.sin(mx**2 + my**2)
```
Create a matrix `array`, `mz`, of `z` values.

```
ax.plot_surface(mx,my,mz)
```
Plot surface of `mx`, `my`, and `mz`.

```
ax.plot_surface(mx,my,mz, cmap=cm.jet)
```
Plot surface of `mx`, `my`, and `mz` with color map.

```
ax.plot_wireframe(mx,my,mz)
```
Plot a wireframe of `mx`, `my`, and `mz`.

Saving Plots

```
fig, ax = plt.subplots(figsize=(3,2))
```
Set the `fig` size to 3″ × 2″ to set text size.

```
plt.savefig('fig1.svg')
```
Save `fig` in a .svg file `fig1.svg`.

```
plt.savefig('fig2.png')
```
Save `fig` in a .png file `fig2.png`.

```
plt.savefig('fig3.png',dpi=600)
```
Save `fig` in a 600 dpi png file `fig3.png`.

LAB PROBLEMS

©=Write only the Python command(s).

®=Write only the Python result.

Solutions to all starred (*) problems are available on the Elsevier website (refer to page xix for the link).

3.1* A common function used in signal processing is the *sinc* function, defined as $\text{sinc}(t) = \sin(t)/t$. Using exactly 200 points, plot the sinc function for $t = -20$ to 20. Use a thick blue line, 10 times thicker than the default value. ©®

3.2 A commonly used function in digital signal processing is the Hamming window and its variants. The equation for this function is

$$w(t) = k - (1-k)\cos\left(\frac{2\pi t}{T}\right) \qquad (3.11)$$

which is plotted over $0 \leq t \leq T$. On a single plot axis, compare the plots for $k = 0.4$, 0.5, and 0.54. Draw each with a different color and different line type (solid, dashed, and dotted) and include a legend. Use a line width of 2 to make the lines easier to see and set $T = 10$ for each plot. Hint: The first step is to create a time vector `array t`. Do this with 100 points. ©®

3.3 Plot the voltage waveform

$$v(t) = 2 + e^{-\frac{t}{5}} \cos(2t) \qquad (3.12)$$

in the style an oscilloscope would show, including a grid, for $0 \leq t \leq 10$ seconds. Label all axes and title the plot. ©®

3.4 A student ill-advisedly places a 1 Ω resistor in a US electrical socket, causing it to dissipate a time-varying power, in watts, of

$$p(t) = \left[120\sqrt{2}\cos(2\pi 60 t)\right]^2 \qquad (3.13)$$

Plot this power for 0.1 seconds using 1000 points. Title the plot "Power Across 1 Ω" and label the horizontal axis "Time, t(s)" and the vertical axis "Power, p(t) (W) ". ©®

3.5* Signals and systems classes teach how to design and analyze filters that remove signals based on their frequency. One such transfer function has the form

$$H(f) = \frac{100}{100 + j2\pi f} \qquad (3.14)$$

where j is the imaginary value $\sqrt{-1}$ and f is the frequency in Hz. $H(f)$ is complex, so it can be viewed in complex polar notation as having a magnitude and an angle.

a. Plot the **magnitude** of $H(f)$ as f varies between 1 and 100 Hz. To do this, use the `np.logspace()` method to create f, and plot the result in a Bode-style plot with the vertical axis in dB and the horizontal axis logarithmically scaled. ©®

b. Repeat the above, but using a linearly-scaled (the normal scale) horizontal axis. ®

3.6 A high-pass filter is a circuit that allows high-frequency signals to pass but blocks slowly changing signals. A transfer function for one type of high-pass filter is given by

$$H(f) = \frac{j20\pi f}{10 + j20\pi f} \qquad (3.15)$$

where j is the imaginary value $\sqrt{-1}$ and f is the frequency in Hz. $H(f)$ is complex, so it can be viewed in complex polar notation as having a magnitude and an angle.

a. Plot the **magnitude** of $H(f)$ as f varies between 0.01 and 100 Hz. To do this, use the `np.logspace()` method to create f, and plot the result in a Bode-style plot with the vertical axis in dB and the horizontal axis logarithmically scaled. Label the figure "High-pass filter". ©®

b. Repeat the above, but using a linearly scaled horizontal axis (the normal scale). ®

3.7 Control classes teach how to analyze systems that could control a robot arm's horizontal motion with varying degrees of damping λ, as described by

$$x(t) = \lambda e^{-\lambda t} \qquad (3.16)$$

Create three sets of line plots on the same set of axes showing the effect of $\lambda = 0.5$, 1, and 2 over the range of $0 \leq t \leq 5$. Use three different linestyles, such as solid, dashed, and dotted. Each plot will, therefore, be readable in a black and white printout. Create the corresponding legend for it. ©®

3.8* Component values are usually within some tolerance of their nominal value. A student takes 10 capacitors from a bin labeled "4.7 µF" and measures the following capacitance values, all in µF: 4.81, 4.23, 5.02, 5.25, 4.86, 4.93, 4.60, 4.85, 4.40, and 4.45. Produce a scatterplot of these values in µF, using points and not connected lines. For example, plot 4.81, not 0.00000481. Use the measurement number, 1−10, for the horizontal axis. Add horizontal lines showing the limit 10% over and under the nominal 4.7 value. Hint: A horizontal line can be drawn with two points. Title the plot "4.7 µF 10% capacitor values". Note that µ is the Greek character mu, µ, not the Latin character u. ©®

3.9 High-tension power lines use high voltages to transmit power so that currents and, therefore, losses from resistance can be kept relatively low. Many typical high-voltage power lines operate at 138 kV. Even so, voltages decrease over the power line with distance, as some power is lost to wire resistance and becomes heat that radiates into the air. A particular power line has measured voltages [138 136 135 134 133 132] kV at distances [0 4 8 12 16 20] km from the generating station. Create a scatterplot using large markers of your choice showing these values. Label the horizontal axis "Distance from generator (in km)" and the vertical axis "Voltage (in kV)", and title the plot "Power Transmission". ©®

3.10 Digital integrated circuits (ICs) have small but unavoidable propagation delays between the time a signal changes at the input and the time the signal changes at the output. These delays are one reason that computer speeds are limited. A student measures the following delays, all in ns (billionths of a second), for different samples of a CMOS chip: 22.9, 24.9, 21.1, 20.2, 20, 22.9, 21.2, 23.1, 22, and 19.9. Scatterplot these values in ns (i.e., you can ignore the "ns") using the measurement numbers 1−10 for the horizontal axis. Place a horizontal line showing the maximum acceptable limit of delay of 24.2 ns (one component falls outside this range). Hint: You can define a horizontal line with two points. Title the plot "Propagation delays in ns". ©®

3.11* Create a plot of a blue square wave $y1$ and a red sine wave $y2$, both with heights ranging from −1 to +1 and extending from $0 \le t \le 4$ as shown in Fig. 3.51. Hints:
- Manually specify eight $\{t1, y1\}$ pairs to make the square wave, but calculate the sine wave using $\{t2, y2\}$ with many points to make it smooth.
- From trigonometry, $\sin(\pi t)$ will have a period $T = 2$, so two cycles will fit in the span of 4 seconds, as shown in Fig. 3.51. ©®

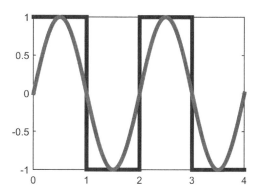

Fig. 3.51 Blue square wave and red sine wave with heights ranging from −1 to +1 and extending from $0 \le t \le 4$ (Lab Problem 3.11).

3.12 The sawtooth waveform shown in Fig. 3.52 is commonly encountered in engineering. It can be approximated by a sum of sine waves given by the equation

$$y(t) = \frac{1}{2} - \frac{\sin(\pi t)}{\pi} - \frac{\sin(2\pi t)}{2\pi} - \frac{\sin(3\pi t)}{3\pi} - \frac{\sin(4\pi t)}{4\pi} \qquad (3.17)$$

Plot Eq. 3.17's approximation to the sawtooth for $0 \le t \le 8$ using 1000 points. Use a second plot command to plot the exact straight-edged sawtooth wave shown in Fig. 3.52. The two images should overlie one another so that it is apparent how closely Eq. 3.17 models the ideal sawtooth. Hint: Unlike the equation's sine wave approximation to the sawtooth, it only takes nine points to plot the ideal straight-line sawtooth. Use a line width of 2 to make the plots more easily visible. ©®

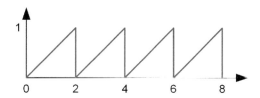

Fig. 3.52 Sawtooth wave with period 2 over the time interval $0 \le t \le 8$ (Lab Problem 3.12).

3.13 Plot the vector of 145 resistor values from 1 Ω to 1 MΩ that you saved in Chapter 2 using `subplots()` to create two plot axes over one another. The top one should have the normal (linear) vertical axis, and the bottom should have a logarithmic vertical axis. Both horizontal axes should count from 1 to 145 for each of the standard 5% resistor values and have normal, linear horizontal axis. Which plot presents the data better? ©®

3.14* Industrial motors and long-distance power transmission lines often use three-phase power. The voltages in one such system are given by the relations

$$v_A(t) = 120 \sqrt{2} \cos{(2\pi 60 t)} \tag{3.18}$$

$$v_B(t) = 120 \sqrt{2} \cos{\left(2\pi 60 t + \frac{2\pi}{3}\right)} \tag{3.19}$$

$$v_C(t) = 120 \sqrt{2} \cos{\left(2\pi 60 t - \frac{2\pi}{3}\right)} \tag{3.20}$$

where the arguments of the cosine function are in radians, not degrees. Plot the three waveforms v_A, v_B, and v_C for $0 \leq t \leq 0.05$ seconds on a single axis. Use 1000 points to plot them, plot each waveform in a different color, and use a legend with subscripts to show which is v_A, v_B, and v_C.

3.15 The complex exponential $f(\omega) = e^{-j\omega}$ is frequently used in signal processing, but it is difficult to graph directly since for every real value of ω, $f(\omega)$ evaluates to a complex value. To visualize it, create two subplots, both in the same row, using `subplots()`. Let ω vary from 0 to 8 using a spacing of 0.1. You cannot name a variable ω in Python, so use w instead, since it is similar.

In the first subplot, create two different line plots on the same set of axes. One should be the real part of $f(\omega)$, and the other should be the imaginary part of $f(\omega)$. Both use the same w values. Use a legend to tell them apart. Label the horizontal axis with the Greek letter ω and give it the title $e^{-j\omega}$ using superscripts.

In the second subplot, plot the real part against the imaginary part. In other words, for this second subplot, use the real part of $e^{-j\omega}$ as the x values and the imaginary part of $e^{-j\omega}$ as the y values, and plot (x, y). Title it "real vs. imaginary part", and label the horizontal and vertical axes "real part" and "imaginary part", respectively. ©®

3.16 Control systems theory can help predict how a system responds to a sudden input change such as flipping a light switch on. This is called the system's step response. Three common types of step response are defined by the relations

$$e^{-\frac{t}{20}} \tag{3.21}$$

$$e^{-\frac{t}{2}} + \frac{t}{2} e^{-\frac{t}{2}} \tag{3.22}$$

$$e^{-\frac{t}{8}} \cos{\left(\frac{t}{5}\right)} \tag{3.23}$$

Plot the three responses for $0 \leq t \leq 75$ seconds using Eqs. 3.21–3.23. Use 1000 points to plot them, plot each equation in a different color, and use a legend to describe them. In the legend, call Eq. 3.21 *overdamped*, Eq. 3.22 *critically damped*, and Eq. 3.23 *underdamped*.

3.17* Plot the three-dimensional function

$$z(x,y) = \frac{2}{e^{(x-0.5)^2+y^2}} - \frac{2}{e^{(x+0.5)^2+y^2}} \tag{3.24}$$

using a grid spanning x and y from -3 to 3 in 0.2 increments. Use the `plot_surface()` command.

3.18 Common lossy image compression algorithms rely on basis functions of the form

$$z(x,y) = \sin(x)\cos(y) \tag{3.25}$$

Plot Eq. 3.25 using a grid spanning x and y from -5 to 5 in 0.2 increments. Use the `plot_wireframe()` command to plot. Title the plot "2D Basis Function" since, although the plot is a 3D plot, it is a basis function of a 2D image. Hint: The plot should look like several smooth hills or an egg crate.

3.19 Using the *IEEE Code of Ethics* set out in this chapter, write a one-paragraph answer to the following case study:

Your undergraduate thesis involves work on signal processing routines that can identify the type of cardiac (heart) disease based on the analysis of electrocardiogram (ECG) signals. Electrocardiogram signals are the voltage signals measured on the chest surface that propagate from the electrical depolarization of the heart as it beats. Manufacturers X and Y go to court over a dispute on patent rights involving a system that uses a related technology, and you are offered a temporary 2-week job to work as an expert witness for Company X. Your job is to explain to the judge why Y's approach infringes on X's patent claims. As you prepare for the case, you write down the many reasons that this is true; however, you also realize there are several reasons that the opposite argument may be made. At the trial, after you deliver your report, the opposing lawyer asks you during cross-examination if you have any reasons to doubt your reported findings. Does your ethical obligation to your client, Company X, outweigh your ethical obligation to completely disclose your doubts? Would your answer change if the opposing expert witness testified first and, when faced with the same line of questioning, answered "of course not", giving the judge a biased view?

3.20 You have been tasked by an automotive company to improve airbag safety by designing a system that either (a) works very well with passengers of common heights but fails to protect passengers who are either unusually tall or short, or (b) works suboptimally for all passengers equally. Which of the 10 articles of the IEEE Code of Ethics described in this chapter provide guidance for this choice? How would you approach the decision, considering the IEEE code?

PYTHON PROGRAMMING

4.1 OBJECTIVES

After completing this chapter, you will be able to use Python to do the following:

- Use *scripts* and *functions* to write structured programs
- Use functions to simplify programs, promote code reuse, and support maintainability
- Define function *input parameters* using *required*, *keyed*, and *default* parameters
- Define a function `return` statement to return an arbitrary number of objects
- Write and `import` *modules* to simplify programs
- Use commenting effectively
- Sort data
- Use *logical operators* to create *relational expressions*
- Use `if`, `if-else`, and `if-elif-else` conditional structures
- Create *formatted strings* with embedded objects

4.2 STRUCTURED PYTHON PROGRAMMING

A Python program is simply a sequence of *definitions* and *commands* that one could *interactively* type during a *session* one by one at the Python *interpreter shell* prompt, >>>. The Python interpreter shell allows one to experiment and test individual lines of code or program statements. These program statements, however, can be grouped together into a text file, typically having a .py extension, or a Jupyter Notebook code cell. All your Python programming, however, could be accomplished by typing commands one by one at the Python interpreter shell prompt. So why program beyond using a Python interpreter shell? A set of program statements assembled into a file or code cell are important because they

1. Allow a sequence of program statements to be saved. After a Python session ends, all variable definitions are lost. Saving these statements in a file, such as a Python text file extension of .py or a Jupyter Notebook file, preserves your work.
2. Hide complexity. For example, a user can execute a program to plot data without having to remember or understand how the program works.
3. Break a larger problem down into smaller, more manageable pieces or subprograms.

4.3 WRITING AND RUNNING PYTHON SCRIPTS

A *script* is a sequence of Python statements that normally produce output when run. You can consider a Python script as a *top-level* program or starting point. In the spirit of other computer languages, such as C++ and Java, the script is referred to as the *main* program.

Consider creating a script to compute the value of two resistors in parallel. The script would need to perform the following:

1. Define the needed variables. For example, define a script to find the equivalent parallel resistance of variables `R1` and `R2`.
2. Calculate the parallel resistance and store it in variable `Rp_R1R2`.
3. Print out the parallel resistance `Rp_R1R2`.

The Python interpreter shell statements and Jupyter Notebook code cell script to perform this task are given in Fig. 4.1.

```
Python Interpreter          >>>        Jupyter Notebook            .ipynb
>>> # Parallel resistor              1  # Parallel resistor
>>> # calculation                    2  # calculation
>>>                                  3
>>> R1 = 10.0                        4  R1 = 10.0
>>> R2 = 15.0                        5  R2 = 15.0
>>> Rp_R1R2 = (R1*R2)/(R1+R2)        6  Rp_R1R2 = (R1*R2)/(R1+R2)
>>>                                  7
>>> print('Rp_R1R2 =',Rp_R1R2)       8  print('Rp_R1R2 =',Rp_R1R2)
Rp_R1R2 = 6.0                           Rp_R1R2 = 6.0
```

Fig. 4.1 Python interpreter statements and Jupyter Notebook code cell script for calculating the resistance of two parallel resistors. Variables for resistors must be predefined with the same names used in the calculation.

Notice that the variables `R1` and `R2`, used to calculate the parallel resistance `Rp_R1R2`, must be predefined with the same names as those used in the calculation of `Rp_R1R2`. There is no flexibility in how the two resistor variables are named. A more complicated script may create intermediate variables to help work out the final answer. Both the output variable that holds the answer and any intermediate variables created to help compute the answer still exist during the session.

PRACTICE PROBLEMS

4.1   Write a script with the variables `V` and `R` defined. It should create a variable called `I` and set it equal to the current defined by Ohm's Law, `V=IR`. ©

Consider the case where two parallel circuits are evaluated in the same script, one with `R1` and `R2` and a second with `R3` and `R4`. One could perform these calculations using the program shown in Fig. 4.2, formed by cutting and pasting the original script and changing the variable names.

You can see how adding three or more parallel resistor calculations starts to increase the program size and the number of variables left over. One of the paradigms of good, structured programming is not to repeat yourself. So there must be a better way. Indeed, there is.

```
Python Interpreter                                                          >>>
>>> # Calculate two sets of parallel resistors
>>>
>>> R1 = 10.0
>>> R2 = 15.0
>>> Rp_R1R2 = (R1*R2)/(R1+R2)
>>>
>>> R3 = 100.0
>>> R4 = 300.0
>>> Rp_R3R4 = (R3*R4)/(R3+R4)
>>>
>>> print('Rp_R1R2 =', Rp_R1R2)
Rp_R1R2 = 6.0
>>> print('Rp_R3R4 =', Rp_R3R4)
Rp_R3R4 = 75.0
```

```
Jupyter Notebook                                                         .ipynb
 1  # Calculate two sets of parallel resistors
 2
 3  R1 = 10.0
 4  R2 = 15.0
 5  Rp_R1R2 = (R1*R2)/(R1+R2)
 6
 7  R3 = 100.0
 8  R4 = 300.0
 9  Rp_R3R4 = (R3*R4)/(R3+R4)
10
11  print('Rp_R1R2 =', Rp_R1R2)
12  print('Rp_R3R4 =', Rp_R3R4)
Rp_R1R2 = 6.0
Rp_R3R4 = 75.0
```

Fig. 4.2 Python interpreter statements and Jupyter Notebook code cell script for calculating the resistances of two sets of parallel resistors. Copying, pasting, and modifying statements to treat different cases increases the number of defined variables and the program complexity.

4.4 INTRODUCTION TO FUNCTIONS

The parallel resistor program has a very specific application that does not promote reusability. If the same program is needed elsewhere, copying and pasting will generally require changing the variable names. In addition, if the original program is modified, changes may have to be made everywhere it is used. This leads to more code maintenance and the possibility of introducing software bugs. The ability to define functions overcomes these limitations. Functions achieve this by defining how variables are passed into and out of the function and how variables created inside the function are handled after the function ends.

You have already used several built-in and imported Python functions. Recall the `exp()` and `sin()` functions imported with the NumPy package. For example,

```
>>> import numpy as np
>>>
>>> y = np.exp(-6)                    # The input variable is -6 and y is set
```

```
>>>                                    # to the function's result
>>> x = np.sin(np.pi)                  # The input variable is np.pi and x is set
>>>                                    # to the function's result
```

These functions will work with inputs other than `-6` or `np.pi` and still give you the correct output.

4.4.1 Function Definitions

The structure of a Python function is outlined in Fig. 4.3. To write your own function, use the `def` keyword followed by the function name and parenthesis to define the input parameters. The function declaration line ends with a colon (`:`), and the next line must begin with an indentation of typically four spaces. Most code editors will automatically add the indentation for you following the carriage return after the colon. A `return` statement exists to optionally return one or more objects to the calling program. A `return` statement with no parameters is the same as returning `None`.

Using the Python interpreter, a function to calculate the resistance of two resistors connected in parallel can be defined as

```
>>> def RparallelFunction(R1,R2):
...        Rparallel = (R1*R2)/(R1+R2)
...        return Rparallel
...
>>>
```

```
Python Interpreter                                              >>>
>>> def functionName(inputParameter1, inputParameter2, ...):
...        # indented body of function
...        return object(s)
...
>>>
```

```
Jupyter Notebook                                            .ipynb
1  def functionName(inputParameter1, inputParameter2, ...):
2      # indented body of function
3      return object(s)
4
```

Fig. 4.3 Python function definition. The keyword `def` begins the function definition. Zero or more input formal parameters are enclosed by round brackets and separated by commas. `None` or any number of objects can be returned to the calling program using the `return` statement at the end of the function.

Using this function, the original program to calculate the parallel resistance of two resistors can be improved as shown in Fig. 4.4. Table 4.1 provides a comparison of the program with and without a function definition.

Notice that the script without a function requires two sets of input variables to be defined, R1 and R2 in the first case and R3 and R4 in the second case. In both cases, an output variable that has a specific name such as Rp_R1R2 or Rp_R3R4 is created. The function, however, does not require input variables to be defined at all. In this example, the variables were passed directly in and passed directly out. One could have defined them using the same names as the function declaration,

```
>>> R1 = 10.0
>>> R2 = 15.0
>>> Rp = RparallelFunction(R1,R2)
```

But the variables also could have had any other names. For example, the set of statements

```
>>> FirstR = 10.0
>>> SecondR = 15.0
>>> output = RparallelFunction(FirstR,SecondR)
```

```
Python Interpreter                                              >>>
>>> # Parallel resistor calculation using a function definition
>>>
>>> def RparallelFunction(R1,R2):
...     Rparallel = (R1*R2)/(R1+R2)
...     return Rparallel
...
>>> Rp = RparallelFunction(10.0, 15.0)
>>> print('Rp Case 1 =', Rp)
Rp Case 1 = 6.0
>>> print('Rp Case 2 =', RparallelFunction(100.0, 300.0))
Rp Case 2 = 75.0
```

```
Jupyter Notebook                                            .ipynb
1  # Parallel resistor calculation using a function definition
2
3  def RparallelFunction(R1,R2):
4      Rparallel = (R1*R2)/(R1+R2)
5      return Rparallel
6
7  Rp = RparallelFunction(10.0, 15.0)
8  print('Rp Case 1 =', Rp)
9  print('Rp Case 2 =', RparallelFunction(100.0, 300.0))
Rp Case 1 = 6.0
Rp Case 2 = 75.0
```

Fig. 4.4 Adding a function definition simplifies the program. Notice how repeated code is avoided when a function is used.

TABLE 4.1   Comparison of parallel resistor scripts with and without a function definition

Script without function	Script with function
```R1 = 10.0R2 = 15.0Rp_R1R2 = (R1*R2)/(R1+R2)R3 = 100.0R4 = 300.0Rp_R3R4 = (R3*R4)/(R3+R4)print('Rp_R1R2 =', Rp_R1R2)print('Rp_R3R4 =', Rp_R3R4)```	```def RparallelFunction(R1, R2): Rparallel = (R1*R2)/(R1+R2) return RparallelRp = RparallelFunction(10.0, 15.0)print('Rp Case 1 =', Rp)print('Rp Case 2 =', RparallelFunction(100.0, 300.0))```
Remaining variables	*Remaining variables*
```R1 = 10.0R2 = 15.0Rp_R1R2 = 6.0R3 = 10.0R4 = 15.0Rp_R3R4 = 75.0```	```Rp```Returns `6.0` to the variable `Rp`The returned value `75.0` is directly printed

will work just as well as those in the previous example. In fact, the input variables do not need to be defined separately. The equivalent example

```
>>> output = RparallelFunction(10.0,15.0)
```

requires no input variables.

For smaller programs, it is convenient to place your function in a separate Jupyter Notebook code cell and run the code cell to define your function. This will make it available to the rest of the Notebook. By the function being placed in its own code cell, it will only need to be run when it is defined or when you make changes to it. Rewriting Fig. 4.4 as two code cells gives us the program shown in Fig. 4.5.

4.4.2 Function Execution

When a function is called, or invoked, from a program, it is given a set of input parameters or arguments. For example, the statement

```
>>> RparallelFunction(10.0,15.0)
```

```
In [1]:
```
```
Jupyter Notebook                                                    .ipynb
  1  # Parallel resistor function definition
  2
  3  def RparallelFunction(R1,R2):
  4      Rparallel = (R1*R2)/(R1+R2)
  5      return Rparallel
```

```
In [2]:
```
```
Jupyter Notebook                                                    .ipynb
  1  Rp = RparallelFunction(10.0, 15.0)
  2  print('Rp Case 1 =', Rp)
  3  print('Rp Case 2 =', RparallelFunction(100.0, 300.0))
```
```
Rp Case 1 = 6.0
Rp Case 2 = 75.0
```

Fig. 4.5 Place your function definition in its own code cell. If you put your function definition in its own code cell and call it from another code cell, you will isolate future modification to one code cell.

calls the function `RparallelFunction` with the input parameters, or arguments, `10.0` and `15.0`. The program execution passes to the first line of the function definition,

```
>>> def RparallelFunction(R1,R2):
```

The first line tells the function to create temporary variables `R1` and `R2` and to assign them the objects that were passed into the function, `10.0` and `15.0`, respectively, in this example. These temporary variables `R1` and `R2` will not remain after the function returns to the calling program. In addition, the variables `R1` and `R2` will not interfere with any other variables in the calling program, even if the calling program already holds declared variables. This is true even if the calling program has variables of the same name as those used in the function program, `R1` and `R2`, for example.

Except for the way variable names are handled, the rest of the function works the same way as the original program. For example,

```
...      Rparallel = (R1*R2)/(R1+R2)
```

creates a temporary variable `Rparallel` and assigns it the result of the computation, with the temporary variables `R1` and `R2`. Here the variable `Rparallel` is not left in memory once the program exits, and if the calling program had a variable named `Rparallel`, the temporary variable would not interfere with it.

The `return` statement

```
...     return Rparallel
```

exits the function program and returns the object `Rparallel` back to the calling program. This can be viewed if the user calls the function in the command line. For example,

```
>>> print(RparallelFunction(10.0, 15.0))
```

The user can set a variable of their choice using an assignment statement such as

```
>>> Rout = RparallelFunction(10.0, 15.0)
```

This creates the variable `Rout` and sets it equal to `6.0`. In either case, the temporary variables `R1`, `R2`, and `Rparallel` defined inside the function will not be left in the calling program's memory. In addition, if the command window had already defined variables `R1`, `R2`, and `Rparallel`, they would not be changed once the function ran.

In summary, functions pass whatever variables they need in and out using the function declaration, eliminating the possibility of variable name conflicts. Any variables created while a function executes will be removed once the function ends, leaving only the variable(s) that the function returns. This behavior of the function promotes structured programming and code reuse.

PRACTICE PROBLEMS

4.2   Clear all variables from Python's memory by exiting and restarting Python if using the interactive shell or by restarting the kernel if using a Jupyter Notebook. Write the script

```
R1 = 10.0
R2 = 15.0
Isource = 5.0
Rp = (R1*R2)/(R1+R2)
V = Isource*Rp
I1 = V/R1
```

This script defines three variables, `R1`, `R2`, and `Isource`, and then computes the current `I1`. What are the names and values of the variables left in memory after the scrip is run? Use the `dir()` command at the Interactive Python prompt or the `%whos` command if using a Jupyter Notebook. Note that the Python `dir()` command will also display a number of other predefined variables unrelated to your script, so do not include them. If you are using the Python interpreter, type the names of the variables at the interpreter prompt `>>>` to find their values. ®

4.3   Rewrite the script in Practice Problem 4.2 using the function

```
def myFunction(R1, R2, Isource):
    Rp = (R1*R2)/(R1+R2)
    V = Isource*Rp
    I1 = V/R1
    return I1
```

Remove all variables from Python's memory, and then run the function as

```
Ioutput = myFunction(10.0, 15.0, 5.0)
```

What are the names and values of all the variables that are in the current Python session; that is, what are the variables that are visible after the function runs? ®

4.5 FUNCTION INPUT PARAMETERS

4.5.1 Zero Input Parameters

Occasionally, a function that does not require any inputs is needed. For example, the program `hello()`, listed in Fig. 4.6, has no input parameters and only returns the `None` object. Note that there is nothing inside the parentheses, since no objects are being assigned to any function parameters. This function is called, or invoked, using `hello()`.

```
Python Interpreter              >>>
>>> # Function definition with
>>> # no inputs or returns
>>>
>>> def hello():
...     print('Hello there!')
...     return None
...
>>> hello()
Hello there!
```

```
Jupyter Notebook              .ipynb
1  # Function definition with
2  # no inputs or returns
3
4  def hello():
5      print('Hello there!')
6      return None
7
8  hello()
Hello there!
```

Fig. 4.6 Function that takes no input parameters and returns the None object. There are no input parameters in the function definition, and None is returned. The return None statement could be left out.

4.5.2 Assigning Input Parameters Using Required Positional Format

The simplest way to assign input arguments to a set of function *formal parameters* is using required positional format. The number and order of the input arguments at the point where the function is called, or invoked, must be the same as the number and order of formal parameters in the function definition. The number of input parameters can be zero, one, or any number, provided the order and number used in the function call are the same as in the function definition.

A function to calculate and print the total number of seconds given seconds, minutes, and hours could be defined and executed as shown in Fig. 4.7. In this example, object 1 is bound to the formal parameter secs, object 2 to mins, and object 3 to hrs. The order and number matter. Two or four input arguments, for example, will cause an error.

4.5.3 One or More Input Parameters Using Key Value Pairs

Using key—value pairs allows the order of the input parameters to be changed. Figure 4.8 demonstrates an application of key—value pairing. Note that when key—value pairs are used, the input argument order in the calling function does not matter. If you combine required positional formatting with key—value pairs, the required position parameters must appear first in the required order before any key—value pairs.

4.5.4 One or More Input Parameters Using Default Values

Using default values in the formal function parameter definitions allows some input arguments to be omitted. Consider the program listing in Fig. 4.9. In the first invocation of the function

```
Python Interpreter                                          >>>
>>> # Calculate total seconds given seconds, minutes, & hours
>>>
>>> def totalSeconds(secs, mins, hrs):
...     totalSecs = secs + 60 * mins + 3600 * hrs
...     print('Total seconds:', totalSecs)
...     return None
...
>>> totalSeconds(3, 2, 1)
Total seconds: 3723
```

```
Jupyter Notebook                                          .ipynb
```

In [1]:
```
1  # Calculate total seconds given seconds, minutes, & hours
2
3  def totalSeconds(secs, mins, hrs):
4      totalSecs = secs + 60 * mins + 3600 * hrs
5      print('Total seconds:', totalSecs)
6      return None
```

In [2]:
```
1  totalSeconds(3, 2, 1)
```
Total seconds: 3723

Fig. 4.7 Assigning input arguments to function formal parameters using required positional format. The number and order of the input arguments (3, 2, 1) must be the same as the number and order of the formal function parameters.

```
Python Interpreter                                          >>>
>>> # Calculate total seconds given seconds, minutes, & hours
>>>
>>> def totalSeconds(secs, mins, hrs):
...     totalSecs = secs + 60 * mins + 3600 * hrs
...     print('Total seconds:', totalSecs)
...     return None
...
>>> totalSeconds(secs=3, mins=2, hrs=1)
Total seconds: 3723
>>> totalSeconds(mins=2, secs=3, hrs=1)
Total seconds: 3723
>>> totalSeconds(hrs=1, secs=3, mins=2)
Total seconds: 3723
```

```
Jupyter Notebook                                          .ipynb
```

In [1]:
```
1  # Calculate total seconds given seconds, minutes, & hours
2
3  def totalSeconds(secs, mins, hrs):
4      totalSecs = secs + 60 * mins + 3600 * hrs
5      print('Total seconds:', totalSecs)
6      return None
```

In [2]:
```
1  totalSeconds(secs=3, mins=2, hrs=1)
2  totalSeconds(mins=2, secs=3, hrs=1)
3  totalSeconds(hrs=1, secs=3, mins=2)
```
Total seconds: 3723
Total seconds: 3723
Total seconds: 3723

Fig. 4.8 Assigning input arguments to formal parameters using key—value pairing. When key—value pairing is used, the order of the input arguments does not matter.

```
>>> total_Seconds(secs=4, mins=2)
```

the `hrs` parameter defaults to a value of `0`. Note that changing the order of the input arguments by invoking the function with

```
>>> total_Seconds(mins=2, secs=4)
```

will not change the result. In the second invocation

```
>>> total_Seconds(mins=2)
```

the `hrs` parameter defaults to a value of `0` and the `secs` parameter to `3`. In the last example no input arguments are given, so all three function parameters default to their values given in the function definition.

```
Python Interpreter                                          >>>
>>> # Calculate total seconds given seconds, minutes, & hours
>>>
>>> def totalSeconds(secs=3, mins=1, hrs=0):
...     totalSecs = secs + 60 * mins + 3600 * hrs
...     print('Total seconds:', totalSecs)
...     return None
...
>>> totalSeconds(secs=4, mins=2)
Total seconds: 124
>>> totalSeconds(mins=2)
Total seconds: 123
>>> totalSeconds()
Total seconds: 63
```

```
Jupyter Notebook                                          .ipynb

In [1]:
1  # Calculate total seconds given seconds, minutes, & hours
2
3  def totalSeconds(secs=3, mins=1, hrs=0):
4      totalSecs = secs + 60 * mins + 3600 * hrs
5      print('Total seconds:', totalSecs)
6      return None

In [2]:
1  totalSeconds(secs=4, mins=2)
2  totalSeconds(mins=2)
3  totalSeconds()
Total seconds: 124
Total seconds: 123
Total seconds: 63
```

Fig. 4.9 Using default parameter values. If the function is called without an input parameter, the parameter is set to its default value.

4.6 FUNCTION RETURN OBJECTS

4.6.1 One Return Object

Functions that return only one object are very common. Examples include the NumPy functions exp() and sin(). A function with input parameters, abbreviated as ..., that returns a single object called output is outlined in Fig. 4.10. For example, the function defined in Fig. 4.11 converts an input value in millimeters to inches and then returns it to the calling program.

```
Python Interpreter          >>>
>>> # Function with one
>>> # return object
>>>
>>> def functionName(...):
...       # body of function
...       return output
...
>>>
```

```
Jupyter Notebook          .ipynb
1  # Function with one
2  # return object
3
4  def functionName(...):
5        # body of function
6        return output
7
8
```

Fig. 4.10 Function with one return object. The return statement at the end of the function is used to return the single object output to the calling program.

```
Python Interpreter          >>>
>>> # Convert mm to inches
>>>
>>> def mm2inches(mm):
...       inches = mm / 25.4
...       return inches
...
>>>
```

```
Jupyter Notebook          .ipynb
1  # Convert mm to inches
2
3  def mm2inches(mm):
4        inches = mm / 25.4
5        return inches
6
7
```

Fig. 4.11 Conversion from mm to inches. An input value in millimeters is assigned to the formal parameter mm, converted to inches, and then returned to the calling program.

Notice that using / means the function mm2inches() will work given either a scalar or an array. The function mm2inches() is called using the syntax

```
>>> myVar = mm2inches(254)
```

After this function is called, myVar holds 10.0. If the function is called without a variable assignment, the returned value is simply printed. For example,

```
>>> mm2inches(254)
10.0
```

4.6.2 Two or More Return Objects

Sometimes more than one object needs to be returned from a function. In this case, each returned object is listed after the function return statement and they are separated by a comma from each other as outlined in Fig. 4.12. An example of this is a program that

```
Python Interpreter                    >>>
>>> # Function with two
>>> # return arguments
>>>
>>> def myFunction(...):
...      # body of function
...      return out1, out2
...
>>>
```

```
Jupyter Notebook                     .ipynb
1  # Function with two
2  # return arguments
3
4  def myFunction(...):
5      # body of function
6      return out1, out2
7
8
```

Fig. 4.12 Function with more than one return object. When more than one return object is needed, the objects follow the return statement and are separated by commas.

converts mm to both inches and feet. The `return` statement includes the two objects inches and feet, separated by a comma.

The function is called setting the returned variables to the new program variables; for example,

```
>>> out1returned, out2returned = myFunction(...)
```

A function that converts millimeters to feet and inches could be written as shown in Fig. 4.13.

```
Python Interpreter                                       >>>
>>> # Convert mm to inches and feet
>>>
>>> import numpy as np
>>>
>>> def mm2ftin(mm):
...      inches = mm / 25.4
...      feet = np.floor(inches/12)
...      inches = inches - feet*12
...      return feet, inches
...
>>> myfeet, myinches = mm2ftin(5000)
>>> print('myfeet:', myfeet)
myfeet: 16.0
>>> print('myinches:', myinches)
myinches: 4.850393700787407
```

```
Jupyter Notebook                                       .ipynb

In [1]:
1  # Convert mm to inches and feet
2
3  import numpy as np
4
5  def mm2ftin(mm):
6      inches = mm / 25.4
7      feet = np.floor(inches/12)
8      inches = inches - feet*12
9      return feet, inches

In [2]:
1  myfeet, myinches = mm2ftin(5000)
2  print('myfeet:', myfeet)
3  print('myinches:', myinches)

myfeet: 16.0
myinches: 4.850393700787407
```

Fig. 4.13 Converting mm to both inches and feet. The input mm is converted to inches and feet, and both objects are returned to the calling program.

RECALL

The `numpy.floor()` function returns the largest integer less than or equal to the argument. `floor(3.5)`, for example, returns `3.0`.

Notice that the function is called using the syntax

```
>>> myfeet, myinches = mm2ftin(5000)
```

Now the variable `myfeet` equals `16.0` and `myinches` equals `4.850393700787407`. Notice another power of using functions. Since you previously created a function `mm2inches()`, you could change the second line in `mm2ftin` from

```
...     inches = mm / 25.4
```

to

```
...     inches = mm2inches(mm)
```

which simplifies your program by calling a function from within your function.

```
Python Interpreter              >>>
>>> # Function definition
>>> # returning None object
>>>
>>> def noneFunction(...):
...         # body of function
...         return None
...
>>>
```

```
Jupyter Notebook            .ipynb
1 | # Function definition
2 | # returning None object
3 |
4 | def noneFunction(...):
5 |         # body of function
6 |         return None
7 |
8 |
```

Fig. 4.14 Function returning the `None` object. If no objects need to be returned, use `None` as the returned object. Omitting the `return None` statement using just `return` has the same meaning as `return None`. It is generally considered good programming practice, however, to explicitly include the `return None` statement in this case.

4.6.3 Returning the None Object

Occasionally, functions do not require any object to be returned, such as when a function just displays a plot. In this case the `None` object is either explicitly or implicitly returned. An example you have already encountered is the `ax.plot()` command. The declaration for a zero-output function is shown in Fig. 4.14.

Note that leaving out the `return` statement, using `return` with no objects, and `return None` are all valid ways of indicating that no explicit values are returned to the calling program. It is, however, good programming practice to explicitly show a `return None` statement. Consider, for example, the function in Fig. 4.15, which creates an artistic-looking plot and returns no objects to the calling program.

```
Python Interpreter                                          >>>
>>> # Generate a cool plot
>>>
>>> import numpy as np
>>> from numpy.fft import fft
>>> import matplotlib.pyplot as plt
>>>
>>> def coolplot():
...     fig, ax = plt.subplots()
...     z = fft(np.eye(11))
...     ax.plot(z.real, z.imag)
...     plt.show()
...     return None
...
>>> coolplot()
```

```
Jupyter Notebook                                          .ipynb
```

```
In [1]:
    1  # Generate a cool plot
    2
    3  import numpy as np
    4  from numpy.fft import fft
    5  import matplotlib.pyplot as plt
    6  %matplotlib inline
    7
    8  def coolplot():
    9      fig, ax = plt.subplots()
   10      z = fft(np.eye(11))
   11      ax.plot(z.real, z.imag)
   12      return None
```

```
In [2]:
    1  coolplot()
```

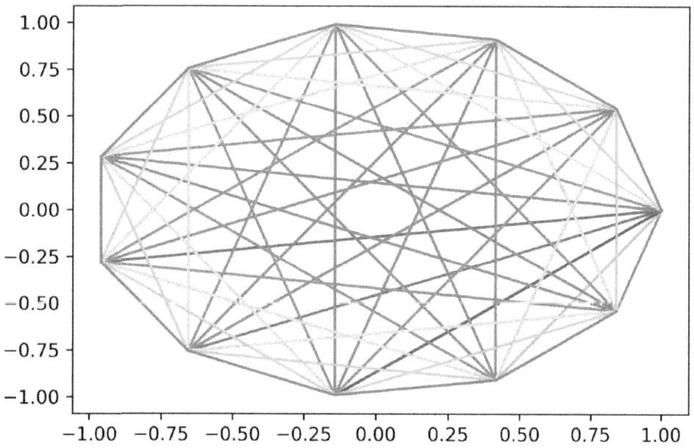

Fig. 4.15 A function that generates a cool plot and returns the None object.

NOTE: PLOTS AS FUNCTIONS
It is often useful to write custom functions to create plots for reports, especially if they are customized from the default plot options. Reports are often revised, and it will save time to be able to quickly regenerate a plot with up-to-date data or to alter its style.

4.6.4 Input and Return Objects Can be Scalars or Arrays

Both input parameters and returned objects can be scalars or arrays. For instance, take a custom simultaneous equation solver that takes the *A* coefficient matrix and the *b* coefficient vector described in Chapter 2 and computes the result using the function given in Fig. 4.16. The inputs `A` and `b` are single matrix and vector arrays, respectively, although both could contain thousands of elements.

```
Python Interpreter                                            >>>
>>> # Solution to Ax = b
>>>
>>> import numpy as np
>>>
>>> def simultaneous(A,b):
...      x = np.linalg.solve(A,b)
...      return x
...
>>>
```

```
Jupyter Notebook                                          .ipynb
1  # Solution to Ax = b
2
3  import numpy as np
4
5  def simultaneous(A,b):
6      x = np.linalg.solve(A,b)
7      return x
```

Fig. 4.16 Input parameters and return objects can be scalars or arrays. The arrays, `A` and `b`, can contain an arbitrary number of elements.

PRACTICE PROBLEMS

4.4 Write the function declaration and `return` statement for a function that computes the maximum frequency of a square wave generated by a microcontroller given the frequency of the microcontroller's clock. That is, write the first and last lines of the function *but not the code that calculates the answer.* Name the function `clock1` and have it take one input parameter, called frequency, and produce one output, called `squarewave`. ©

4.5 Write the function declaration and `return` statement for a function that takes one vector of values, such as a `NumPy vector array` of all standard 5% resistor values, and a second input that describes how many random resistor values are requested and returns a

NumPy vector array of that many random resistors. Only write the function declaration and return statement, not the code to solve the problem. Call the function problem4_5. ©

4.6.5 Passing Immutable and Mutable Parameters

When objects are passed to a function, they are bound to *formal parameters* in a process called *pass by assignment*. Recall that formal parameters are declared in the first line of the function as comma separated variables; for example,

```
>>> def functionName(input1, input2, ...):
```

If the object is immutable, that is, cannot be changed, the variable can be bound to a new object, but the original object cannot be changed. Python cannot change the value of an immutable object by reassigning the formal parameter to a new value in a function. Once the function ends, the original object prior to the calling of the function remains unchanged. Consider a function that accepts an input parameter in degrees and calculates the equivalent value in radians as shown in Fig. 4.17. Notice how the variable outside the function is not changed.

If the object passed to a function is mutable, however, it can be altered in a function. Mutable objects, such as NumPy arrays, cannot be altered outside of the function by assignment, but they can be mutated. Changing the elements of an array, for example, will persist after the function has ended. The program listing in Fig. 4.18 alters a single element of a NumPy array.

4.7 CREATING AND IMPORTING MODULES WITH FUNCTIONS

4.7.1 Introduction

A *module* is like a script, but it is imported into another program or Python interpreter session to increase its functionality. As your programs increase in size and complexity, it makes sense to save your functions and other definitions in module files so they can be imported into other programs. A module is written as a text file, usually with the ending .py. A package is a folder containing one or more modules and a special module called __init__ that initializes the package. You have already used the import statement to import modules and packages

```
Python Interpreter                                                    >>>
>>> # Degrees to radians conversion program
>>>
>>> import numpy as np
>>>
>>> def degToRads(x):
...     print('x inside degToRads just after entry:', x)
...     x = x * np.pi / 180.0 # convert from degrees to rads
...     print('x inside degToRads just after conversion:', x)
...     return None
...
>>> x = 90.0
>>> degToRads(x)
x inside degToRads just after entry: 90.0
x inside degToRads just after conversion: 1.5707963267948966
>>> print('x after returning from degToRads:', x)
x after returning from degToRads: 90.0
```

In [1]:
```
Jupyter Notebook                                                  .ipynb
1  # Degrees to radians conversion program
2
3  import numpy as np
4
5  def degToRads(x):
6      print('x inside degToRads just after entry:', x)
7      x = x * np.pi / 180.0 # convert from degrees to rads
8      print('x inside degToRads just after conversion:', x)
9      return None
```

In [2]:
```
1  x = 90.0
2  degToRads(x)
3  print('x after returning from degToRads:', x)
x inside degToRads just after entry: 90.0
x inside degToRads just after conversion: 1.5707963267948966
x after returning from degToRads: 90.0
```

Fig. 4.17 Degrees to radians conversion. When immutable objects, such as numerical objects, are passed to a function, the original value remains unchanged outside the function.

already. The NumPy package, for example, was imported to allow your programs to manipulate data arrays.

4.7.2 Writing Module Text Files

To create your own module, use a text editor, such as Notepad, Notepad++, nano, Sublime Text, Emacs, or Vim, to create the text file Rparallel.py in your current directory. Do not use Word or another word processor that does not save your file in a text format. The function to calculate the parallel resistance of two resistors can be written in a separated text file as a module, Rparallel.py. The text file Rparallel.py can be imported in the interactive interpreter, >>>, or a Jupyter Notebook code cell using the import command, as illustrated in Fig. 4.19.

```
Python Interpreter                                        >>>
>>> # Modifying array values in a function
>>>
>>> import numpy as np
>>>
>>> def add4ToIndex1(x):
...         print('x inside add4ToIndex1 just after entry:', x)
...         x[1] = x[1] + 4 # add 4 to index 1 value
...         print('x inside add4ToIndex1 just after adding:', x)
...         return None
...
>>> x = np.array([1, 2, 3])
>>> add4ToIndex1(x)
x inside add4ToIndex1 just after entry: [1 2 3]
x inside add4ToIndex1 just after adding: [1 6 3]
>>> print('x after calling add4ToIndex1:', x)
x after calling add4ToIndex1: [1 6 3]
```

```
Jupyter Notebook                                        .ipynb

In [1]:
    1  # Modifying array values in a function
    2
    3  import numpy as np
    4
    5  def add4ToIndex1(x):
    6      print('x inside add4ToIndex1 just after entry:', x)
    7      x[1] = x[1] + 4
    8      print('x inside add4ToIndex1 just after adding:', x)
    9      return None

In [2]:
    1  x = np.array([1, 2, 3])
    2  add4ToIndex1(x)
    3  print('x after calling add4ToIndex1:', x)
x inside add4Toindex1 just after entry: [1 2 3]
x inside add4Toindex1 just after adding: [1 6 3]
x after calling add4Toindex1: [1 6 3]
```

Fig. 4.18 Altering a mutable object such as a NumPy array in a function. Some objects, such as NumPy arrays, are mutable, and changes made within a function persist after the function returns to the calling program.

4.8 CREATING YOUR OWN FUNCTION SUMMARY

There are several steps for creating your own functions:

1. Decide the names of the input parameters, since the function will access these parameters within the function. Use this information to write the function declaration, which is the first line of the function.
2. Then decide what commands one would type to solve the problem using the input parameter names. These commands form the remainder of the function. As part of this section, consider if the input parameters will only be scalars or could be vector or matrix arrays. Check that any outputs required by the calling program follow the function return statement at the end of the function.

```
Python Interpreter              >>>
>>> # Importing a module
>>>
>>> import Rparallel as Rpm
>>>
>>> Rp = Rpm.RpFunction(10,15)
>>> print('Rp =', Rp )
Rp = 6.0
```

```
Jupyter Notebook            .ipynb
1  # Importing a module
2
3  import Rparallel as Rpm
4
5  Rp = Rpm.RpFunction(10,15)
6  print('Rp =', Rp)
Rp = 6.0
```

```
Python                   Rparallel.py
# Parallel Resistor Module

def RpFunction(R1,R2):
    Rparallel=(R1*R2)/(R1+R2)
    return Rparallel
```

Fig. 4.19 Importing modules into a Python program. A module is a text file containing Python code. Using a text editor, Python code is written in a text file ending in .py that can be imported into another program. Use the `import` statement to give your program access to the definitions in the module file. In this case, the module Rparallel is imported with the alias Rpm. To use the function defined in this file, use the prefix Rpm followed by a period.

Example

An earlier practice problem involved a script to use Ohm's Law, $V = IR$, to solve for the current given a voltage and a resistance.

Solution

Input parameters `V` and `R`

Output argument `I`

Function declaration

```
>>> def ISolveFunction(V, R):
```

The command using only the arguments is `I = V/R`. Note that using `/` allows either or both `V` and `R` to be `arrays`.

Check that the output variable `I` is assigned.

In summary, to solve the problem, the text outlined in Fig. 4.20 should be entered into a Jupyter Notebook code cell or written into a text file and imported as a module.

```
Python Interpreter                                          >>>
>>> # Summary function using Ohm's law to find current
>>>
>>> def ISolveFunction(V,R):
...     I = V/R
...     return I
...
>>> I = ISolveFunction(10,5)
```

```
Jupyter Notebook                                          .ipynb
```

```
In [1]:
    1  # Summary function using Ohm's law to find current
    2
    3  def ISolveFunction(V,R):
    4      I = V/R
    5      return I
```

```
In [2]:
    1  I = ISolveFunction(10,5)
```

Fig. 4.20 Summary function to find current given voltage and resistance. The current, I, is calculated using Ohm's Law from the input parameters V and R. The current is then returned to the calling program.

PRACTICE PROBLEMS

4.6   Create a function called `seriesResistance` that takes two resistor values and returns their equivalent series resistance, that is, their sum. ©

4.7   Create a function called `parallelResistance` that takes two resistor values and returns their equivalent parallel resistance, that is, their product divided by their sum. ©

4.9 CREATING HELP WITH DOCSTRINGS

The pound sign # is used in Python to denote a comment. All information on the line following a # symbol is ignored. Use this to insert comments to help you or someone else understand what your program does or to help break very large programs into blocks. Continuing from the example shown in Fig. 4.20, Fig. 4.21 illustrates the use of additional informative comments. For short programs, such as this one, you may not see the need for commenting. As your programs grow to tens or hundreds of lines of code, however, comments become very important.

```
Python Interpreter          >>>     Jupyter Notebook          .ipynb
>>> # Commenting programs          1  # Commenting programs
>>>                                 2
>>> def ISolveFunction(V, R):       3  def ISolveFunction(V, R):
...     I = V/R # Ohm's law         4      I = V/R # Ohm's law
...     # return current            5      # return current
...     return I                    6      return I
...                                 7
>>>                                 8
```

Fig. 4.21 Using the # character to add comments. Everything after the # character is ignored.

Python functions have documentation that can be displayed using the Python `help()` command. Use the name of the function you wish to display documentation for by using the function name as an input parameter of the `help()` function. In Fig. 4.22 the `help()` command displays the documentation associated with the built-in function `len()` and the `NumPy` `linspace()` function.

```
Python Interpreter          >>>     Jupyter Notebook          .ipynb
>>> # Using help with built-in      1  # Using help with built-in
>>> # and imported functions        2  # and imported functions
>>>                                 3
>>> help(len)                       4  help(len)
>>> import numpy as np              5  import numpy as np
>>> help(np.linspace)               6  help(np.linspace)
```

Fig. 4.22 Using the `help()` function to display documentation. The `help()` function displays the documentation associated with the function name given as the `help` argument.

You can build help documentation into your functions as well. This is done using Python documentation strings, or *docstrings*. Docstrings are declared using triple single quotation marks, `'''`, or triple double quotation marks, `"""`, just below the function declaration. It is good practice to provide a docstring with all functions. For a simple program, a one-line docstring will suffice as illustrated in Fig. 4.23.

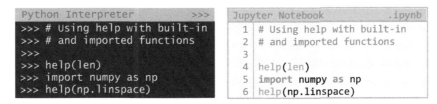

```
Python Interpreter          >>>     Jupyter Notebook          .ipynb
>>> # Using a one-line              1  # Using a one-line
>>> # docstring                     2  # docstring
>>>                                 3
>>> def square(x):                  4  def square(x):
...     """Return x squared"""      5      """Return x squared
...     return x*x                  6      return x*x
...                                 7
>>>                                 8
```

Fig. 4.23 Using a one-line docstring in a function. For small functions, a one-line docstring will be adequate.

For more complicated programs, multiline docstrings are used. Some conventions for multiline docstrings include the following:

1. The docstring line should be in sentence structure beginning with a capital letter and ending with a period.
2. The first line is a short description.
3. If there are more lines in the documentation string, the second line should be blank, visually separating the summary from the rest of the description.
4. The following lines should be one or more paragraphs describing the object's calling convention and so on.

The basic format for a multiline docstring based on these conventions is outlined in Fig. 4.24. Applied to the previous function, a Python docstring is shown in Fig. 4.25.

```
Python Interpreter                                              >>>
>>> # Creating docstrings
>>>
>>> def myFunction(inputArg):
...     """
...     Summary line
...
...     Detailed function description
...
...     Input parameter(s)
...         inputArg: Description of inputArg
...
...     Return object(s)
...         Description of return object(s)
...
...     """
...
```

```
Jupyter Notebook                                           .ipynb
 1  # Creating docstrings
 2
 3  def myFunction(inputArg):
 4      """
 5      Summary line
 6
 7      Detailed function description
 8
 9      Input parameter(s)
10          inputArg: Description of inputArg
11
12      Return object(s)
13          Description of returned object(s)
14
15      """
```

Fig. 4.24 Creating docstrings. Docstrings help document your functions. Using the `help()` function with the name of your function as an argument will display the docstring.

```
Python Interpreter                                              >>>
>>> # Using docstrings in function definitions
>>>
>>> def ISolveFunction(V, R):
...     """
...         Finds the current I given voltage V and resistance R.
...
...         This function calculates and returns the current given
...             the voltage and resistance using Ohms's law.
...
...         Input parameters
...             V: Voltage in Volts
...             R: Resistance in Ohms
...
...         Return object
...             I: Current in Amperes
...
...     """
...
...     I = V/R # Calculate the current based on Ohm's Law
...
...     return I
...
>>>
```

```
Jupyter Notebook                                              .ipynb
 1  # Using docstrings in function definitions
 2
 3  def ISolveFunction(V, R):
 4      """
 5          Finds the current I given voltage V and resistance R.
 6
 7          This function calculates and returns the current given
 8              the voltage and resistance using Ohm's law.
 9
10          Input parameters
11              V: Voltage in Volts
12              R: Resistance in Ohms
13
14          Return object
15              I: Current in Amperes
16      """
17
18      I = V/R # Calculate the current based on Ohm's law
19
20      return I
```

Fig. 4.25 Using a docstring in an Ohm's Law current calculation function.

When the user types

```
>>> help(ISolveFunction)
```

The help command will return

```
Help on function ISolveFunction in module __main__:

ISolveFunction(V, R)
    Finds the current I given voltage V and resistance R.
    This function calculates and returns the current given
        the voltage and resistance using Ohm's law.
```

```
Input parameters
    V: Voltage in Volts
    R: Resistance in Ohms

Return object
    I: Current in Amps
```

The same information will be returned using the command

```
>>> print(ISolveFunction.__doc__)
```

Notice the extra blank lines added inside the program. Python ignores blank lines, so they can help organize your program into logical blocks such as the help and code areas in this example.

PRACTICE PROBLEMS

4.8 Create a commented function `F2C()` that converts a temperature in Fahrenheit to a temperature in Celsius. It should run using the command `F2C(32)`, for example, which will return 0. The conversion formula is $C = (5/9) (F-32)$. ©
The function should implement help. Typing

```
>>> help(F2C)
```

should return the message

```
Help on function F2C in module __main__:

F2C(F)
    Returns a temperature in Celsius given one in Fahrenheit.

    This function reads in a temperature F in Fahrenheit and
    uses the relation C = (5/9) (F-32) to convert to the
    temperature C in Celsius and returns it.

    Input parameter
        F: Input temperature in Fahrenheit
    Return object
        C: Output temperature in Celsius
```

4.10 MORE COMPLEX FUNCTION EXAMPLES

Examine the following functions to develop your programming ability.

4.10.1 Example: Converting a Complex Number From Polar Radian to Cartesian Form

Complex numbers in electrical engineering are often given in complex polar form, written with a magnitude and an angle that can be in radians or degrees. These were introduced in Chapter 2 in section 2.6, *Complex Numbers*. An example is $23\angle0.7854$ (radians) or $23\angle45°$ (degrees). Python requires complex numbers to be entered in rectangular form, using real parts and imaginary parts such as `4+3j` or `6*np.cos(np.pi/4)-1j*np.sin(np.pi/3)`. Chapter 2 described how a complex number can be changed from polar $mag\angle angle$ form into rectangular `realpart+1j* imagpart` form using

```
>>> import numpy as np
>>> realpart = mag * np.cos(angle)
>>> imagpart = mag * np.sin(angle)
```

Once the real and imaginary parts of the number are obtained, they can be assembled into a single complex number `z` using the command

```
>>> z = realpart + 1j * imagpart
```

Note that the `NumPy` functions `cos()` and `sin()` expect angles in radians. If you need to convert to angles in degrees, use the `NumPy` function `deg2rad()`, for example, `cos(deg2rad())`. A function to do the entire conversion is provided in Fig. 4.26.

To use it to input $23\angle0.5$ radians into variable `x`, call it using the syntax

```
>>> x = polarRadian2Complex(23, 0.5)
```

To four decimal places, the variable x will have the value `x = 20.1844 + 11.0268j`.

>
> **NOTE**
> Inside the function, the return variable is called "z," but it was called a completely different name, "x," by the calling program. This is one of the strengths of using a function. The person using the function does not have to know what the function calls its variables internally.

```
Python Interpreter                                                    >>>
>>> # Polar radian to Cartesian complex number conversion
>>>
>>> import numpy as np
>>>
>>> def polarRadian2Complex(magnitude, phase):
...     """
...     Converts a complex number from polar to Cartesian form.
...
...     Converts a complex number represented in polar form
...         with a magnitude and phase into a complex number
...         represented in Cartesian form with real and
...         imaginary parts.
...
...     Input parameters
...         magnitude: Polar representation magnitude.
...         phase:     Polar representation phase in radians.
...
...     Return object
...         z: Cartesian complex number with real and
...             imaginary parts.
...     """
...     realpart = magnitude * np.cos(phase)
...     imagpart = magnitude * np.sin(phase)
...     z = realpart + imagpart * 1.0j
...     return z
...
>>>
```

```
Jupyter Notebook                                                  .ipynb
 1   # Polar radian to Cartesian complex number conversion
 2
 3   import numpy as np
 4
 5   def polarRadian2Complex(magnitude, phase):
 6       """
 7       Converts a complex number from polar to Cartesian form.
 8
 9       Converts a complex number represented in polar form
10           with a magnitude and phase into a complex number
11           represented in Cartesian form with real and
12           imaginary parts.
13
14       Input parameters
15           magnitude: Polar representation magnitude.
16           phase:     Polar representation phase in radians.
17
18       Return object
19           z: Cartesian complex number with real and
20               imaginary parts.
21
22       """
23       realpart = magnitude * np.cos(phase)
24       imagpart = magnitude * np.sin(phase)
25       z = realpart + imagpart * 1.0j
26       return z
```

Fig. 4.26 Complex number conversion from polar to Cartesian form.

4.10.2 Example: Low-Pass Filter Magnitude Bode Plot for a Given Cutoff Frequency

Low-pass filters were introduced in the Tech Tip *Bode Plots* in Section 3.2.9. Courses in signals and systems will derive the first-order low-pass filter equation

$$H(f) = \frac{1}{1 + j\left(\dfrac{f}{f_c}\right)} \tag{4.1}$$

where f is the input frequency in Hz, f_c the cutoff frequency in Hz, and $H(f)$ *is* called the frequency response. A low-pass filter is designed to pass signals with frequencies below f_c and attenuate those with frequencies above f_c. Note that $H(f)$ is a complex function of the input frequency f since it has a j term in the denominator.

>
> **RECALL**
> A Bode-style plot uses horizontal frequency values on log base 10 scale and plots the vertical amplitude axis in dB. To find the value in dB of a complex number z, use the relation $20 \log_{10}|z|$, where $|z|$ denotes the absolute value of the complex number z.

Problem

Create a function called lowpass() that takes a cutoff frequency f_c and creates a Bode plot of the magnitude of $|H(f)|$ for $1 \leq f \leq 100$ Hz.

Solution

Clear all your old variables and output. If you are using a Python shell, exit using the `exit()` command and restart it by typing python at the window command prompt. If you are using a Jupyter Notebook, choose "Restart & Clear Output" under the Kernel drop-down menu.

First `import` the needed modules and packages for numerical analysis and plotting:

```
>>> import numpy as np                    # needed for numerical analysis
>>> import matplotlib.pyplot as plt       # needed for plotting
>>> #%matplotlib inline                   # only used in a Jupyter Notebook
```

Since the function takes one number (f_c), the function definition line and comments are

```
>>> def lowpass(fc):
...        """
...        lowpass(fc=10) creates a low pass filter magnitude plot.
...        """
```

Then create a frequency vector array, logarithmically spaced between 1 and 100 Hz, using

```
...        f = np.logspace(0, 2, 50) # 1-100Hz = 10**0 - 10**2
```

Compute the frequency response, $H(f)$, and the amplitude in dB:

```
...        H = 1.0/(1.0+1.0j*(f/fc))
...        dB = 20*np.log10(np.abs(H))
```

Complete the plot and label it using the statements

```
...        fig, ax = plt.subplots()
...        ax.plot(f,dB)
...        ax.set_xlabel('Frequency, f (Hz)')
...        ax.set_xscale('log')
...        ax.set_ylabel('Magnitude, |H(f)| (dB)')
...        ax.grid(which='both', color='b', ls='-.', lw=0.25)
...        plt.show()
```

No objects are returned to the calling program, so use the optional `return None` statement to make this clear:

```
...        return None
...
>>>
```

Call the program to create a 10 Hz low-pass filter using

```
>>> lowpass(10)
```

The Bode-style plot shown in Fig. 4.27 is displayed.

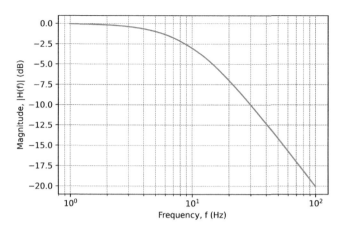

Fig. 4.27 Low-pass 10 Hz cutoff filter magnitude Bode plot. Frequencies near 1 Hz are passed, while higher frequencies are attenuated.

Note that none of the variables created inside the function remain after the function returns. The complete documented programs using the Python interpreter and a Jupyter Notebook are listed in Fig. 4.28.

```
Python Interpreter                                      >>>
>>> # Low pass filter Bode plot
>>>
>>> import numpy as np
>>> import matplotlib.pyplot as plt
>>>
>>> def lowpass(fc):
...     """
...     Bode plot generation
...
...     Creates a Bode-style plot of the low pass filter
...         frequency response with cutoff frequency fc.
...
...     Input parameter
...         fc: Cutoff frequency in Hz
...
...     Return object
...         None
...
...     """
...     # Create a frequency array log spaced from 1-100 Hz
...     f = np.logspace(0, 2, 50) # 1-100 Hz = 10**0 - 10**2
...     # Compute the frequency response and amplitude in dB
...     H = 1.0/(1.0+1.0j*(f/fc))
...     dB = 20.0 * np.log10(np.abs(H))
...     # Complete the plot and label it
...     fig, ax = plt.subplots()
...     ax.plot(f,dB)
...     ax.set_xlabel('Frequency, f (Hz)')
...     ax.set_xscale('log')
...     ax.set_ylabel('Magnitude, |H(f)| (dB)')
...     ax.grid(which='both', color='b', ls='-.', lw=0.25)
...     plt.show()
...     # No objects are returned to the calling program
...     return None
...
>>> lowpass(10) # Call the program with fc=10
```

Fig. 4.28a Low-pass filter Bode plot using the Python interpreter. Given an input frequency, *fc*, in Hz, a Bode plot is generated.

Introduction to Python and Spice for Electrical and Computer Engineers

```
Jupyter Notebook                                          .ipynb
  1  # Low pass filter Bode plot
  2
  3  import numpy as np
  4  import matplotlib.pyplot as plt
  5  %matplotlib inline
  6
  7  def lowpass(fc):
  8      """
  9      Bode plot generation
 10
 11      Creates a Bode-style plot of the low pass filter
 12          frequency response with cutoff frequency fc.
 13
 14      Input parameter
 15          fc: Cutoff frequency in Hz
 16
 17      Return object
 18          None
 19      """
 20      # Create a frequency array log spaced from 1-100 Hz
 21      f = np.logspace(0, 2, 50) # 1-100 Hz = 10**0 - 10**2
 22      # Compute the frequency response and amplitude in dB
 23      H = 1.0/(1.0+1.0j*(f/fc))
 24      dB = 20.0 * np.log10(np.abs(H))
 25      # Complete the plot and label it
 26      fig, ax = plt.subplots()
 27      ax.plot(f,dB)
 28      ax.set_xlabel('Frequency, f (Hz)')
 29      ax.set_xscale('log')
 30      ax.set_ylabel('Magnitude, |H(f)| (dB)')
 31      ax.grid(which='both', color='b', ls='-.', lw=0.25)
 32      # No objects are returned to the calling program
 33      return None
```

Fig. 4.28b Low-pass filter Bode plot using a Jupyter Notebook code cell. Given an input frequency, fc, in Hz, a Bode plot is generated.

PRACTICE PROBLEMS

4.9 Modify the Bode plot function to return the filter response in dB rather than plotting it. For example, using $f_c{=}10$, a vector array of 50 values should be returned, the first of which is -0.0432. ©

4.11 SORTING

Sorting `NumPy array` data is an important feature of many application programs. Fortunately, `NumPy arrays` can be sorted easily. The `array` can be sorted with the data values ascending or descending. In addition, it is possible either to `sort` the original `array` or to create a new sorted `array` leaving the original `array` unchanged. To ascending `sort` the original `NumPy` array, use the `.sort()` method appended to the end of the `array` variable name. If you need a descending sort, apply the NumPy `flip()` function as illustrated in Fig. 4.29. The print statements show how the original `array` is sorted.

```
Python Interpreter                                                    >>>
>>> # Sorting numpy arrays
>>> # The .sort() method sorts the original array
>>>
>>> import numpy as np
>>>
>>> oldArray = np.array([4, 9, 1, -6, 0])
>>> print('Original array:     ', oldArray)
Original array:      [ 4  9  1 -6  0]
>>> oldArray.sort()
>>> print('Ascending sorted: ', oldArray)
Ascending sorted:    [-6  0  1  4  9]
>>> oldArray = np.flip(oldArray)
>>> print('Descending sorted: ', oldArray)
Descending sorted:  [ 9  4  1  0 -6]
```

```
Jupyter Notebook                                                   .ipynb
 1  # Sorting numpy arrays
 2  # The .sort() method sorts the original array
 3
 4  import numpy as np
 5
 6  oldArray = np.array([4, 9, 1, -6, 0])
 7  print('Original array      ', oldArray)
 8  oldArray.sort()
 9  print('Ascending sorted:  ', oldArray)
10  oldArray = np.flip(oldArray)
11  print('Descending sorted: ', oldArray)
Original array:      [ 4  9  1 -6  0]
Ascending sorted:    [-6  0  1  4  9]
Descending sorted:  [ 9  4  1  0 -6]
```

Fig. 4.29 Sorting NumPy vector arrays. Using the `.sort()` method replaces the original array with an ascended sorted version of itself. Use the NumPy `flip()` function if you need a descending sort.

If you need the original array and want to create a new sorted array from it, use the NumPy `sort()` function with the original array as an input argument. The NumPy `sort()` function returns a new sorted array while leaving the original array unchanged. The program listing in Fig. 4.30, for example, shows that the newly created array is a sorted version of the original unchanged array. This capability can be useful inside functions that simulate circuit designs over many different random component value variations and will be used in one of the lab problems. If you need the data sorted in descending order instead of ascending order, use the NumPy `flip()` method to reverse the order of the array elements.

```
Python Interpreter                                          >>>
>>> # Creating a new sorted NumPy array
>>> # without changing the original array
>>>
>>> import numpy as np
>>>
>>> oldArray = np.array([4, 9, 1, -6, 0])
>>> newAscendingArray = np.sort(oldArray)
>>> newDescendingArray = np.flip(newAscendingArray)
>>> print('oldArray:           ', oldArray)
oldArray:            [ 4  9  1 -6  0]
>>> print('newAscendingArray: ', newAscendingArray)
newAscendingArray:  [-6  0  1  4  9]
>>> print('newDescendingArray:', newDescendingArray)
newDescendingArray: [ 9  4  1  0 -6]
```

```
Jupyter Notebook                                          .ipynb
 1  # Creating a new sorted NumPy array
 2  # without changing the original array
 3
 4  import numpy as np
 5
 6  oldArray = np.array([4, 9, 1, -6, 0])
 7  newAscendingArray = np.sort(oldArray)
 8  newDescendingArray = np.flip(newAscendingArray)
 9  print('oldArray:           ', oldArray)
10  print('newAscendingArray: ', newAscendingArray)
11  print('newDescendingArray:', newDescendingArray)

oldArray:           [ 4  9  1 -6  0]
newAscendingArray:  [-6  0  1  4  9]
newDescendingArray: [ 9  4  1  0 -6]
```

Fig. 4.30 Creating a new sorting array while leaving the original array unchanged. Using the `NumPy sort()` function with the original array as an input argument leaves the original `NumPy` array unchanged and returns an ascending sorted version of it. To create a descending sorted array, use the `NumPy flip()` function.

PRO TIP: GRADUATE STUDIES IN EE

Master's

There are two types of master's degree programs for electrical engineers: *industrial* and *research*. This is confusing, since there is no standardized name for either type of program. Some universities offer both, while others offer only one of the two. Some call one degree a Master of Science (M.S.) and the other a Master of Engineering (M.Eng.), while some programs call both an M.S. degree. Examining the graduation requirements will make it clear which course(s) a particular university offers.

- *Industrial M.S.*: Industrial M.S. programs prepare students for jobs in industry. They typically require 8—10 classes. A thesis, if required, is small, the equivalent of one or two courses. This takes typically one full year to complete, with the associated cost of tuition.

- **Research M.S.**: This prepares the student for a research position and is the first step in earning a doctorate. Coursework varies widely, from as few as three courses to as many as required for an industrial master's. A thesis is mandatory and is a major component of the work. This takes longer than an industrial master's, often 18–24 months. This is often not a separate degree program but part of a doctoral degree program, so tuition cost may be waived in return for work as a teaching assistant (TA) and research assistant (RA), described below.

Doctorate

Engineering doctoral degree programs are usually sponsored fully through a mix of teaching assistantships (TAs) and research assistantships (RAs), which typically provide tuition remission and a modest stipend to pay for food and basic accommodations.

In the first year, the student usually works as a TA and teaches undergraduate students. During this year, the doctoral candidate is expected to seek longer-term funding from professors as an RA. This is done by visiting research labs whose work the student finds appealing. If the student and the principal investigator, or PI, the laboratory's lead professor, think they may be a good match (i.e., the PI believes the student will be an asset to the laboratory, and the student enjoys the type of research work, likes the other RAs, and trusts the professor), then the student begins working part-time in the laboratory as a trial run. After a semester of this, it should be clear if the lab is a good fit for the student, and the PI will offer an RA to the student. This will allow the student to drop the TA, keep the tuition waiver, and now be paid to do the research that is required by his or her thesis. The PI becomes the student's primary research advisor.

A less common, but more prestigious, type of funding is a fellowship. A doctoral student holding a fellowship receives tuition remission and a stipend comparable to that of an RA, but the source of funding for the fellowship comes from an outside agency, rather than from the PI who funds the RA. This makes students with fellowships very desirable for PIs, who no longer have to pay for their tuition and stipend.

4.12 RELATIONAL OPERATORS

A logical expression is a statement that evaluates to either "true" or "false." Relational operators are a type of logical expression that compare two values such as $5 > 4$ (true) or $3 \leq -4$ (false). Python returns `True` to indicate true and `False` to indicate false. Python has several types of relational operators. Some of the most common relational operators are listed in Table 4.2.

TABLE 4.2 Python relational operators

Operator	Name	Example	Example result
>	Greater than	5 > 2	True
<	Less than	7 < -6	False
>=	Greater than or equal to	a = -5 a >= 6	False
<=	Less than or equal to	a = 7 b = 9 a * b <= a + b	False
==	Equal to	5 == 5	True
!=	Not equal to	5 != 5	False
==	Equal to with strings	a = 'hello' b = 'Hello' a == b	False (capitalization matters)

When applied to arrays, these operators evaluate corresponding elements similar to the operators `+`, `-`, `*`, `/`. Try it. Note that one of the most common relational statements, the test to see if two scalar numbers are the same, is not `=` but rather `==`. An application of the `==` operator is given in Fig. 4.31. The function called `password()` tests to see if the input is "secret" and returns `True` if it is and `False` otherwise.

```
Python Interpreter                                         >>>
>>> # Password test function
>>>
>>> def password(inputStr)
...     """
...     Password test to see if inputStr
...         is the string 'secret'.
...     """
...     result = (inputStr=='secret')
...     return result
...
>>>
```

```
Jupyter Notebook                                         .ipynb
1  # Password test function
2
3  def password(inputStr):
4      """
5      Password test to see if inputStr
6          is the string 'secret'.
7      """
8      result = (inputStr=='secret')
9      return result
```

Fig. 4.31 Password test function. The inputStr parameter is tested to see if it is equal to the string "secret". A `True` or `False` result is returned based on this test.

PRACTICE PROBLEMS

4.10   Modify the function shown in Fig. 4.31 to take a number as an argument and test to see if it is 999. If it is, return `True`; otherwise, return `False`. Call the function `problem4_10()`. ©

4.13 LOGICAL OPERATORS

Logical operators operate on logical true or false arguments and are particularly important to ECE students, since binary true/false signals form the basis of all digital computers. Common logic operators are listed in Table 4.3.

TABLE 4.3   Python logical operators

Operator	Name	Example	Example result
`and`	AND. `True` only if *both* operands are `True`.	x = 5 (x > 1) and (x < 4)	`False` (only one is `True`)
`or`	OR. `True` if *either* or *both* operands are `True`.	x = 5 (x > 1) or (x < 4)	`True` (the first test is `True`)
`not`	NOT. Changes `True` to `False` and `False` to `True`.	a = 'Hello' not a == 'T'	`True` (the strings are `not` equal)

If you are familiar with digital gates, the associated digital logic symbols are shown in Table 4.4.

TABLE 4.4   Digital logic symbols

Operator	Symbol
and	
or	
not	

As an example, suppose you need to check the validity of numbers in a large data set of electronically sampled location measurements. These numbers should all be nonnegative integers less than 10 or could be -1 if the sensor returned an error condition. The function given in Fig. 4.32 tests whether a particular number meets these criteria.

```
Python Interpreter                                          >>>
>>> # Multiple test criteria function
>>>
>>> import numpy as np
>>>
>>> def test(num)
...     """
...     Tests if num is a non-negative integer, <10, or =-1.
...     """
...     isError = (num==-1) # False if non-error, True if error
...     isInteger = (num==np.floor(num)) # True if integer
...     isInRange = (num>=0) and (num<10)  # True if 0<=num<10
...     out = (isInteger and isInRange) or isError
...     return out
...
>>>
```

```
Jupyter Notebook                                          .ipynb
 1  # Multiple test criteria function
 2
 3  import numpy as np
 4
 5  def test(num):
 6      """
 7      Tests if num is a non-negative integer, <10, or =-1.
 8      """
 9      isError = (num==-1)   # False if non-error, True if error
10      isInteger = (num==np.floor(num))   # True if integer
11      isInRange = (num>=0) and (num<10) # True if 0<=num<10
12      out = (isInteger and isInRange) or isError
13      return out
```

Fig. 4.32 Multiple test criteria function. Using relational and logical operators, the function tests the input variable to see if it is non-negative, less than 10, or equal to -1.

4.11 Create a function called `problem4_11()` that takes a scalar, that is, a single number, and returns `True` if the input is not an integer and is negative. For example, calling `problem4_11()` with π and $-\pi$ should return `False` and `True`, respectively. That is,

```
>>> import numpy as np
>>> problem4_11(np.pi)
False
>>> problem4_11(-np.pi)
True ©
```

4.13.1 Logical Array Operations

It is frequently desirable to perform logical operations on large dataset arrays. A logical operation on an array returns an array of logical values. For example, the program listed in Fig. 4.33 tests each element of the array *v* and returns an array of logical values.

```
Python Interpreter          >>>
>>> # Logical array operations
>>>
>>> import numpy as np
>>>
>>> v=np.array([-2,-1,0,1,2])
>>> b=(v>=0)
>>> print(b)
[False False True True True]
```

```
Jupyter Notebook          .ipynb
1  # Logical array operations
2
3  import numpy as np
4
5  v=np.array([-2,-1,0,1,2])
6  b=(v>=0)
7  print(b)
[False False True True True]
```

Fig. 4.33 Logical array operations. Each element of the array is tested to see if it is non-negative.

There are four common actions done with the resulting array of logical values:

- Determine if any of the logical results are `True`
- Determine if all of the logical results are `True`
- Determine how many of the results are `True`
- Replace the values of the array for which the given relationship is `True` with another number

Determine if Any of the Logical Results Are True

Use the `any()` logical operator. In the above example

```
>>> any(b)
```

or equivalently

```
>>> any(v>=0)
```

evaluates to `True`, that is, returns `True`, since some of the values in vector `b` are `True`.

Determine if All of the Logical Results Are True
Use the `all()` logical operator. In the above example

```
>>> all(b)
```

or equivalently

```
>>> all(v>=0)
```

evaluates to `False`, that is, returns `False`, since some of the values in vector `v` are `False`.

Determine How Many Results Are True
Summing the result returns the total number of operations that evaluate to `True`. Using the above example,

```
>>> sum(b)
```

or more directly

```
>>> sum(v>=0)
```

will return `3`.

Replacing Array Values for Which a Given Relationship Is True with Another Number
Replacing array values with another number based on a `True` relation is a very common requirement and has its own syntax. Up to this point, one-dimensional arrays, or vectors, have been indexed with a number, such as `v[4]`, and two-dimensional arrays, or matrices, have been indexed with a pair of numbers, such as `m[3,4]`. However, arrays can also be indexed by a logical array of the same size, as shown in Fig. 4.34. This program replaces negative elements in the vector array `v` with `9`s. That is, where `v<0` evaluates to `True`, the number is replaced by `9`.

```
Python Interpreter            >>>        Jupyter Notebook           .ipynb
>>> # Replacing array values              1  # Replacing array values
>>> # using logical operations           2  # using logical operations
>>>                                       3
>>> import numpy as np                    4  import numpy as np
>>>                                       5
>>> v=np.array([-2,-1,0,1,2])            6  v=np.array([-2,-1,0,1,2])
>>> v[v<0] = 9                            7  v[v<0] = 9
>>> print(v)                              8  print(v)
[9 9 0 1 2]                                  [9 9 0 1 2]
```

Fig. 4.34 Replacing array values with numbers based on logical relations. In this example, whenever an array value is less than zero it is replaced with the value nine.

PRACTICE PROBLEMS

4.12 Create a function called `problem4_12()` that takes an array of numbers and returns two variables. The first return variable, called "count," is the number of values in the input array greater than 10. The second is an array called "result" that is equal to the input but with all values greater than 10 replaced with 10. Use the .copy() method as shown so you do not change the original array x. The function definition should be of the form

```
def problem4_12(x):
    result = x.copy()
    # statement(s)
    return count, result
```

and as an example of its use,

```
import numpy as np
x = np.array([-5, 20, 6, 11])
count, result = problem4_12(x)
```

gives `count=2` and `result=np.array([-5, 10, 6, 10])`. ©

4.14 CONDITIONAL BRANCHING

4.14.1 *if*

Thus far, every Python program discussed has executed every line of code in it when run. The `if` statement block, however, is an exception. The indented code block following the `if` statement is only run when the `if` statement's logical expression is `True`. The `if` statement structure is of the form

```
>>> if (logical expression that evaluates to True or False):
...        # Statements here execute if the logical expression is true.
...        # All statements until the end of the indented block are
...        # conditionally run.
...
>>>
```

For example, the function listed in Fig. 4.35 duplicates `abs(x)`, which returns the absolute value of x. Notice that the statement(s) associated with an `if` block are indented. Python uses indentation to define code blocks. It also makes the code easier to read.

```
Python Interpreter          >>>
>>> # absolute value using if
>>>
>>> def abs_if(x):
...        if (x<0): # x negative
...            x = -x
...        return x
...
>>>
```

```
Jupyter Notebook            .ipynb
1  # absolute value using if
2
3  def abs_if(x):
4      if (x<0): # x negative
5          x = -x
6      return x
7
8
```

Fig. 4.35 Using the `if` statement to implement the absolute value function. The `if` statement tests a logical statement to determine whether the following indented code is executed. In this case, when x is less then zero, the indented code is executed and x is replaced with -x. If x is not negative, the indented code block is not executed, and x is returned to the calling program.

A more complicated example evaluates the `sinc(x)` function, a very common function used in signal processing, defined as

$$\operatorname{sinc}(x) = \begin{cases} \dfrac{\sin(x)}{x}, & x \neq 0 \\ 1, & x = 0 \end{cases} \tag{4.2}$$

The sinc function is implemented using the code listed in Fig. 4.36.

```
Python Interpreter          >>>
>>> # Sinc function
>>>
>>> import numpy as np
>>>
>>> def sinc(x):
...        if x != 0:
...            y = np.sin(x)/x
...        if x == 0:
...            y = 1
...        return y
...
>>>
```

```
Jupyter Notebook            .ipynb
1  # Sinc function
2
3  import numpy as np
4
5  def sinc(x):
6      if x != 0:
7          y = np.sin(x)/x
8      if x == 0:
9          y = 1
10     return y
11
12
```

Fig. 4.36 Using the `if` statement to implement the sinc function. Notice that when x is zero the sinc function must be set to 1 because a division by zero error will occur. The `if` statement avoids a division by zero error.

Some programmers use flowcharts to show how their code flows. While this technique is no longer as common as it used to be, flowcharts make clear the way that program execution branches around `if` blocks. Figure 4.37 shows the flowchart for an `if` code block.

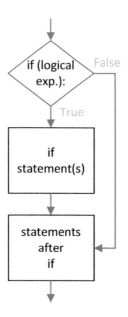

Fig. 4.37 Python `if` statement flowchart. If the logical expression in the `if` statement evaluates to `True`, the indented `if` block statements are executed. Otherwise, the indented `if` statement block is bypassed.

PRACTICE PROBLEMS

4.13 Create a function called `problem4_13()` that takes a string and returns a string set to the phrase "unlocked" if the supplied password is "secret" and "locked" otherwise.

© Hints:
- Python uses both single and double quotation marks to denote strings.
- `==` and `!=` compare numbers, arrays, and strings.

4.14.2 if-else

In the previous section, a `sinc` function was developed that ran one `if` code block when x equaled zero and used a second `if` code block when x was not zero. This need, to run one set of code if the logical test is true, and a different set of code if the logical test is false, is so common that it has its own syntax: `if-else`. The `if-else` conditional branch has the form

```
>>> if (logical expression that evaluates to Ture or False):
...         # Statements here execute if the logical expression is true.
...         # All statements until the end of the indented block are
...         # conditionally run.
... else:
...         # Statements here are executed if the logical expression
...         # is False.
...
>>>
```

The `sinc` function, developed in the previous section, can be simplified using the `if-else` syntax as demonstrated in Fig. 4.38.

```
Python Interpreter                    >>>
>>> # Sinc function
>>>
>>> import numpy as np
>>>
>>> def sinc(x):
...         if x == 0:
...             y = 1
...         else:
...             y = np.sin(x)/x
...         return y
...
>>>
```

```
Jupyter Notebook                    .ipynb
 1  # Sinc function
 2
 3  import numpy as np
 4
 5  def sinc(x):
 6      if x == 0:
 7          y = 1
 8      else:
 9          y = np.sin(x)/x
10      return y
11
12
```

Fig. 4.38 Using the Python `if-else` statement in a sinc function.

All conditionally executing blocks of code can have many statements that run if the logical test is `True` or `False`, although this example has just one line of code in each. Remember to indent the code blocks. The flowchart for the `if-else` is shown in Fig. 4.39.

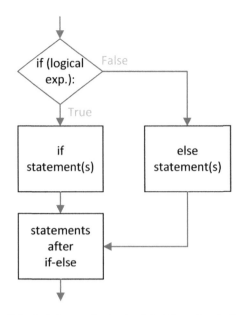

Fig. 4.39 Python `if-else` flow chart. If the logical expression in the `if` statement evaluates to `True`, the body of the indented `if` statements is executed. Otherwise, the body of the indented `else` statements are executed.

PRACTICE PROBLEMS

4.14 Update the previous function, `problem4_13()`, to use the `if-else` statement and save it as `problem4_14()`. ©

4.14.3 if-elif-else

If there is one outcome of a test that requires special code, use an `if` block. If there are two different outcomes that each requires special handing, use an `if-else` block. If there are more than two outcomes that require different code blocks to be executed, the `if-elif-elif-...-elif` block works. Use as many `elif` statements as needed for each possible outcome. A final optional `else` statement can be used to handle all remaining cases. Figure 4.40 shows the `if-elif-else` flowchart. As Figure 4.41 illustrates, increasing the number of `elif` code blocks increases the number of possible outcomes.

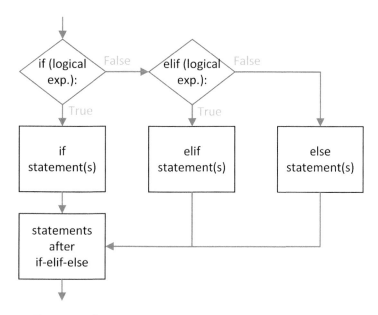

Fig. 4.40 Python `if-elif-else` conditional branching flow chart.

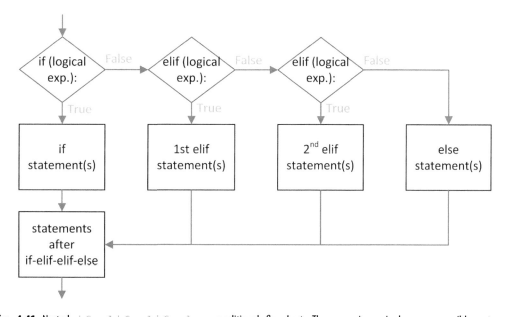

Fig. 4.41 Nested `if-elif-elif-else` conditional flowchart. There are increasingly many possible outcomes associated as the `if-elif-...-else` tests increase.

As an example, one could generate a random number between 0 and 1 and, based on that number, provide different responses. The function listed in Fig. 4.42 prints different responses based on the value of a random number.

```
Python Interpreter                                          >>>
>>> # Is ECE fun:  if-elif-elif-elif-else structure
>>>
>>> import numpy as np
>>>
>>> def isECEfun():
...     """
...     Is ECE fun?
...     """
...     r = np.random.rand() # Random number between 0 and 1
...     if (r<0.2):
...         out = 'always'
...     elif (r<0.4):
...         out = 'sometimes'
...     elif (r<0.6):
...         out = 'usually but not on Mondays'
...     elif (r<0.8):
...         out = 'up until 1:30 AM'
...     else:
...         out = 'can I change my major to biology?'
...     return out
...
>>>
```

```
Jupyter Notebook                                          .ipynb
 1  # Is ECE fun:   if-elif-elif-elif-else structure
 2
 3  import numpy as np
 4
 5  def isECEfun():
 6      """
 7      Is ECE fun?
 8      """
 9      r = np.random.rand()  # Random number between 0 and 1
10      if (r<0.2):
11          out = 'always'
12      elif (r<0.4):
13          out = 'sometimes'
14      elif (r<0.6):
15          out = 'usually but not on Mondays'
16      elif (r<0.8):
17          out = 'up until 1:30 AM'
18      else:
19          out = 'can I change my major to biology?'
20      return out
```

Fig. 4.42 Using the `if-elif-elif-elif-else` construction.

PRACTICE PROBLEMS

4.15 Create a function called `problem4_15()` that computes the sign
      function used in control theory. It takes a number and returns -1
      if the input is negative, 0 if the input is 0, and 1 if the input is
      positive. Use an `if-elif-else` construction. ©

4.15 CREATING FORMATTED STRINGS

4.15.1 Using the str() Function and the String Concatenation Operator +

To create a string with both text and embedded numbers, use the built-in function `str()` to convert the number to a string and the `+` operator to concatenate strings. For example, the program listed in Fig. 4.43 creates a string `currentString` with content `'The current is 36.5 Amps.'` and prints it out.

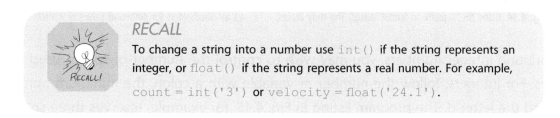

Fig. 4.43 Using the `str()` function and the concatenation operator, `+`, to format strings. The `str()` function converts a number to a string and the `+` operator between two strings concatenates them.

> **RECALL**
> To change a string into a number use `int()` if the string represents an integer, or `float()` if the string represents a real number. For example,
> `count = int('3')` or `velocity = float('24.1')`.

4.15.2 String Formatting Methods

In some cases, you will need more control over how numbers are formatted in your string. For example, it may be necessary to control the number of characters, including white spaces, and the number of digits of precision. There are three methods commonly used to do this: the `f`-prefix, the `str.format()` function, and the `%` operator. The `f`-prefix is the most recent method and is available with Python versions 3.6 and up. The `str.format()` method is available in earlier Python versions. A description of the `%` operator method has been included but is considered out of date and has been known to introduce errors in some cases. It has been included in case you run into earlier Python code samples, particularly those earlier than version 2.6. The choice will depend on your personal preference and application.

Using the Formatted String Literal or f-Prefix

A *formatted string literal*, *f-string*, or *f*-prefix string is a string prefixed with "f" or "F" and uses curly braces {} to define replacement fields for formatted input numbers or other strings. You can specify the number or variable in the braces as demonstrated in Fig. 4.44. Notice how the curly braces act as a placeholder.

```
Python Interpreter                                          >>>
>>> import numpy as np
>>>
>>> VrmsString = f'The rms value is {120} Vrms.'
>>> print(VrmsString)
The rms value is 120 Vrms.
>>> VampValue = np.sqrt(2.0) * 120
>>> VampString = f'The amplitude is {VampValue} Vamp.'
>>> print(VampString)
The amplitude is 169.7056274847714 Vamp.
```

```
Jupyter Notebook                                          .ipynb
1   import numpy as np
2
3   VrmsString = f'The rms value is {120} Vrms.'
4   print(VrmsString)
5   VampValue = np.sqrt(2.0) * 120
6   VampString = f'The amplitude is {VampValue} Vamp.'
7   print(VampString)

The rms value is 120 Vrms.
The amplitude is 169.7056274847714 Vamp.
```

Fig. 4.44 Using the f-prefix to format strings. The curly braces, { }, act as placeholders for numerical values or variables.

When printing integer numbers, you may wish to control the number of spaces used for the numbers. For integers, follow the number or variable with a colon, the minimum number of digits, and the letter d. The program listing in Fig. 4.45, for example, reserves three spaces for the first print statement and six for the second.

```
Python Interpreter                                          >>>
>>> VampValue = 170
>>> VppString1=f'Peak-to-peak amplitude: {2*VampValue:3d} Vpp.'
>>> print(VppString1)
Peak-to-peak amplitude: 340 Vpp.
>>> VppString2=f'Peak-to-peak amplitude: {2*VampValue:6d} Vpp.'
>>> print(VppString2)
Peak-to-peak amplitude:    340 Vpp.
```

```
Jupyter Notebook                                          .ipynb
1   VampValue = 170
2   VppString1=f'Peak-to-peak amplitude: {2*VampValue:3d} Vpp.'
3   print(VppString1)
4   VppString2=f'Peak-to-peak amplitude: {2*VampValue:6d} Vpp.'
5   print(VppString2)

Peak-to-peak amplitude: 340 Vpp.
Peak-to-peak amplitude:    340 Vpp.
```

Fig. 4.45 Using the f-prefix to format a string with an embedded integer. Follow the variable name in the curly brackets with a colon, the minimum number of digits, and the letter d.

Formatting a floating-point number follows a similar pattern. Follow the number or variable with a colon, the minimum number of characters for the number, a decimal point, the number of decimal places, and the letter f. Note that the total number of characters includes the decimal point. For example, the code listing in Fig. 4.46 displays the rms values with two different formats. To use scientific notation, use e or E instead of f, as shown in Fig. 4.47.

```
Python Interpreter                                          >>>
>>> import numpy as np
>>>
>>> VampValue = 170
>>> VrmsValue = VampValue / np.sqrt(2.0)
>>> VrmsString3 = f'The rms value is {VrmsValue:7.3f} Vrms.'
>>> print(VrmsString3)
The rms value is 120.208 Vrms.
>>> VrmsString4 = f'The rms value is {VrmsValue:8.1f} Vrms.'
>>> print(VrmsString4)
The rms value is     120.2 Vrms.
```

```
Jupyter Notebook                                          .ipynb
1  import numpy as np
2
3  VampValue = 170
4  VrmsValue = VampValue / np.sqrt(2.0)
5  VrmsString3 = f'The rms value is {VrmsValue:7.3f} Vrms.'
6  print(VrmsString3)
7  VrmsString4 = f'The rms value is {VrmsValue:8.1f} Vrms.'
8  print(VrmsString4)

The rms value is 120.208 Vrms.
The rms value is     120.2 Vrms.
```

Fig. 4.46 Using the f-prefix to format a string with an embedded floating-point number. Follow the variable name in the curly brackets with a colon, the minimum number of digits, a period, the number of digits after the decimal points, and the letter f.

```
Python Interpreter                                          >>>
>>> import numpy as np
>>>
>>> VampValue = 170
>>> VrmsValue = VampValue / np.sqrt(2.0)
>>> VrmsSciString = f'The rms value is {VrmsValue:9.3e} Vrms.'
>>> print(VrmsSciString)
The rms value is 1.202e+02 Vrms.
```

```
Jupyter Notebook                                          .ipynb
1  import numpy as np
2
3  VampValue = 170
4  VrmsValue = VampValue / np.sqrt(2.0)
5  VrmsSciString = f'The rms value is {VrmsValue:9.3e} Vrms.'
6  print(VrmsSciString)

The rms value is 1.202e+02 Vrms.
```

Fig. 4.47 Using the f-prefix to format a string with an embedded scientific number. Follow the variable name in the curly brackets with a colon, the minimum number of digits, a period, the number of digits after the decimal points, and the letter e or E.

Strings can also be embedded using the s symbol. In the example shown in Fig. 4.48 a minimum of five spaces are reserved for the string. The period shows that within the five spaces the actual string is left justified by default. Additional justification control characters, (<, ^, >), can be used to left, center, and right justify the integer if the minimum number of characters exceeds the number needed to display the integer. The examples shown in Fig. 4.49 illustrate the use of left, center, and right justification characters.

```
Python Interpreter                                              >>>
>>> R = 2200
>>> units = 'Ohms'
>>> RString = f'{R:4d} {units:5s}.'
>>> print(RString)
2200 Ohms .
```

```
Jupyter Notebook                                              .ipynb
1  R = 2200
2  units = 'Ohms'
3  RString = f'{R:4d} {units:5s}.'
4  print(RString)
2200 Ohms .
```

Fig. 4.48 Using the f-prefix to format a string with another embedded and formatted string. Follow the variable name in the curly brackets with a colon, the minimum number of characters, and then the character s.

```
Python Interpreter                                              >>>
>>> x = 21
>>> intString = f'*{x:<6d}* Left justification.'
>>> print(intString)
*21    * Left justification.
>>> print(f'*{x:^6d}* Center justification.')
*  21  * Center justification.
>>> print(f'*{x:>6d}* Right justification.')
*    21* Right justification.
```

```
Jupyter Notebook                                              .ipynb
1  x = 21
2  intString = f'*{x:<6d}* Left justification.'
3  print(intString)
4  print(f'*{x:^6d}* Center justification.')
5  print(f'*{x:>6d}* Right justification.')
*21    * Left justification.
*  21  * Center justification.
*    21* Right justification.
```

Fig. 4.49 Using the f-prefix to format a number within a string with left, center, and right justification. Add a <, ^, or > immediately following the colon to left, center, or right justify the output, respectively.

Using the format() Function to Create Strings with Embedded Numbers

The Python format() function can also be used to format numbers in strings with Python versions 2.6 and up. The formatting is similar to that of the f-string, but the colon is removed

```
Python Interpreter                                              >>>
>>> import numpy as np
>>>
>>> Vstr = 'The AC voltage is '
>>> Vamp = 170
>>> Vrms = Vamp / np.sqrt(2.0)
>>> Str = '{:s}{:3d} Vamp, {:6.3f} Vrms.'.format(Vstr,Vamp,Vrms)
>>> print(Str)
The AC voltage is 170 Vamp, 120.208 Vrms.
```

```
Jupyter Notebook                                             .ipynb
  1  import numpy as np
  2
  3  Vstr = 'The AC voltage is '
  4  Vamp = 170
  5  Vrms = Vamp / np.sqrt(2.0)
  6  Str = '{:s}{:3d} Vamp, {:6.3f} Vrms.'.format(Vstr,Vamp,Vrms)
  7  print(Str)

The AC voltage is 170 Vamp, 120.208 Vrms.
```

Fig. 4.50 Using `format()` to create string with formatted numbers and strings. This is similar in nature to the f-string method, but the values to be formatted are included at the end of the string with a period and as arguments to the `format()` function.

and the values are included by appending a `.format()` at the end of the string. Figure 4.50 demonstrates the use of the .format() function.

Using the String Formatting or Interpolation Operator %

The interpolation operator uses a % symbol as placeholder. The string is followed by a tuple beginning with a % symbol with the variables or values to be formatted. Figure 4.51 illustrates the use of the % command.

```
Python Interpreter                                              >>>
>>> import numpy as np
>>>
>>> Vstr = 'The AC voltage is'
>>> Vamp = 170
>>> Vrms = Vamp / np.sqrt(2.0)
>>> VString = '%s %3d Vamp or %7.3f Vrms.' % (Vstr,Vamp,Vrms)
>>> print(VString)
The AC voltage is 170 Vamp or 120.208 Vrms.
```

```
Jupyter Notebook                                             .ipynb
  1  import numpy as np
  2
  3  Vstr = 'The AC voltage is'
  4  Vamp = 170
  5  Vrms = Vamp / np.sqrt(2.0)
  6  VString = '%s %3d Vamp or %7.3f Vrms.' % (Vstr,Vamp,Vrms)
  7  print(VString)

The AC voltage is 170 Vamp or 120.208 Vrms.
```

Fig. 4.51 Using the % command to format numbers and strings. The % command designates where the formatted string or number goes and how it is formatted. At the end of the string, the variables or values to be formatted are included in round brackets, preceded by the % command.

4.16 Given a resistor variable R, use the number-to-string conversion function `str()` and the concatenation operator `+` to print the statement. © For example, if $R = 26.2$, the expected output will be

The resistance is 26.2 ohms

4.17 Write the % operator, f-prefix, and `format()` based commands that, given a variable R, print the statement to two decimal places. For example, if $R = 26.2146$ the expected output will be

The resistance is 26.21 ohms ©

TECH TIP: WIRE SIZING

Wire sizes in North America are specified by the American Wire Gauge (AWG) standard. Smaller wire diameters have higher wire gauge sizes. General-purpose hookup wire used in electronics prototyping is typically 22 AWG, thicker wire used to carry electricity to kitchen appliances is often 18 AWG, and heavy-duty power extension cables are 14 AWG. Wire size is a function of maximum required current-carrying capacity and of the length over which a wire is required to carry the current. Table 4.5 lists common AWG sizes with wire diameters and their recommended current capacity over short distances, such as inside appliances.

TABLE 4.5 Common American Wire Gauge (AWG) sizes

AWG	Conductor diameter (in.)	Max current capacity (amps)
12	0.0808	41
14	0.0641	32
16	0.0508	22
18	0.0403	16
20	0.0320	11
22	0.0253	7
24	0.0201	4

Wire is also available in two types: solid and stranded. Solid wire is composed of a single, solid (usually copper) conductor. It is preferred for use in prototyping because it can be pushed into the holes of solderless prototyping breadboards, and for power distribution in buildings because it is easier to connect to receptacles. Stranded wire is composed of several strands of smaller wires bundled together. Stranded wire is more flexible than solid-core wire and is preferred in applications that may repeatedly bend, such as the power wires connecting to a laptop or an electric iron.

PRACTICE PROBLEMS

4.18 Write a function called `problem4_18()` that takes an AWG wire size (even sizes 12–24) and returns the following string (using as an example an AWG of 12) "AWG 12 is 0.0808 inches diameter, max current 41 A." Use `if-elif-elif` statements to check the wire size and to set temporary wire diameter and current capacity variables. Then, at the end, use one `format()` statement to create a string using the temporary wire diameter and current capacity variables that are returned by the function. ©

TECH TIP: ENGINEERING NOTATION

Engineers use unit postfixes, letters following numbers, to express large or small numbers so that the mantissas are always between 1 and 1000. This is *not* scientific notation. For example, 4.3×10^{-11} in scientific notation is written in engineering notation as 43 p. These postfixes are in ratios of 1000. Common ones used in our profession are shown in Table 4.6. Note: Capitalization matters!

A 1 MΩ resistor is extremely large; a 1 mΩ resistor is very small. Notice how the numerical parts of all the examples are between 1 and 1000.

TABLE 4.6   Common engineering postfix notations

Prefix	Name	Scaling	Engineering example	Scientific notation (do not use)
p	pico	10^{-12}	12 pF	12×10^{-12} F
n	nano	10^{-9}	−434 nV	-434×10^{-9} V
μ	micro	10^{-6}	1.2 μA	1.2×10^{-6} A
m	milli	10^{-3}	150 mH	0.15H
—	—	1	36 H	36H
k	kilo	10^{3}	13 kV	13,000 V
M	mega	10^{6}	1 MΩ	10^{6} Ω
G	giga	10^{9}	96 GB	96×10^{9} B
T	tera	10^{12}	2 TB	2×10^{12} B

As you get used to using these symbols, you will find they will help you do mathematics spanning large intervals easily. For instance, a 12 μA current through a 4 MΩ resistor generates their product in voltage, which is 12×4 in units of μ, two up from the center, times M, two down from the center, which is simply 48 times the center, or 48 V. No powers of 10 are required!

The `if-elif-else` construct can be used to change numbers into engineering form. An implementation that works for milli, (unity), and kilo prefixes is given in Fig. 4.52.

```
Python Interpreter                                          >>>
>>> # Rewriting numbers in engineering form
>>> def engform(x):
...     """
...     engform takes the input number x and rewrites it in
...     engineering form and returns it to the calling program.
...     For example:
...         0.0032 is returned as 3.2m
...         12452 is returned as 12.452k
...         98.6 is returned as 98.6
...     """
...     if (x<1):        # Value is in milli unit
...         x = x * 1000
...         prefix = 'm'
...     elif (x<1e3):    # No prefix needed if 1 <= x < 1000
...         prefix = ''
...     else:            # Value is in kilo unit
...         x = x / 1000
...         prefix = 'k'
...     s = str(x) + prefix
...     return s
...
>>>
```

```
Jupyter Notebook                                          .ipynb
 1   # Rewriting numbers in engineering form
 2   def engform(x):
 3       """
 4       engform takes the input number x and rewrites it in
 5       engineering form and returns it to the calling program.
 6       For example:
 7           0.0032 is returned as 3.2m
 8           12452 is returned as 12.452k
 9           98.6 is returned as 98.6
10       """
11       if (x<1):        # Value is in milli unit
12           x = x * 1000
13           prefix = 'm'
14       elif (x<1e3):    # No prefix needed if 1 <= x < 1000
15           prefix = ''
16       else:            # Value is in kilo unit
17           x = x / 1000
18           prefix = 'k'
19       s = str(x) + prefix
20       return s
```

Fig. 4.52 Rewriting numbers in engineering form. Using an `if-elif-else` construction, milli and kilo units are added, depending on the value of the number `x`.

TECH TIP: RESISTOR COLOR CODES

The size of a resistor is not related to its resistance, but rather to how much power it can dissipate. Many modern resistors are in extremely small surface mount packages. The 0201 package, for example, is about 0.02″ x 0.01″. Two could fit on the period at the end of this sentence. However, most resistors used for prototyping in solderless breadboards are 1/4 W or 1/8 W sized and look like the magnified image shown in Fig. 4.53. These resistors have four bands of color that, with some practice, can be read by sight according to the diagram and table shown in Fig. 4.54.

Fig. 4.53 Representative prototyping resistor. The color banding gives the resistor value and tolerance.

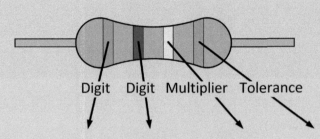

Digit Digit Multiplier Tolerance

Color	Digit	Digit	Multiplier	Tolerance
Black	0	0	1	
Brown	1	1	10	1%
Red	2	2	100	
Orange	3	3	10^3	
Yellow	4	4	10^4	
Green	5	5	10^5	
Blue	6	6	10^6	
Violet	7	7		0.1%
Gray	8	8		
White	9	9		
Gold				5%

Fig. 4.54 Resistor color band code. The above resistor is gray, red, yellow, and gold, meaning $82 \times 10^4 \; \Omega = 820 \; k\Omega$, 5% tolerance.

PRACTICE PROBLEMS

4.19 Write a function that takes a resistor digit from 0 to 9 and returns its color code. Use the `if-elif-else` construct. For example, `s = problem4_19(4)` sets `s` equal to `'Yellow'`. ©

Even at the start of their studies, EE students often wonder if they should enter industry or graduate school after completing their bachelor's degree. It is a hard decision and perhaps best summarized by listing the benefits of each approach.

The benefits of going directly to graduate school:

- Many careers require a graduate degree, such as managing large engineering projects or performing cutting-edge research.
- Many electrical engineers eventually earn a graduate degree, so it makes sense to do so when
  - Material learned as an undergraduate is the freshest
  - Additional life responsibilities are less likely to impinge on studies
  - An early start may help boost career advancement significantly

The benefits of going into industry first:

- Graduate school requires a specialization within EE, and it can be difficult to choose without some industry exposure first.
- Since graduate school admission favors a student's most recent accomplishments, students without high undergraduate GPAs may do well in industry and then enter graduate programs they would not have been accepted into directly out of college.
- New graduates can earn and save significantly during a few years in industry.
- Many employers, especially large ones, pay their employees to get an M.S. in a related field.
- New graduates may find they enjoy the industry position they land with their B.S., thus saving the time, expense, and opportunity cost of pursuing unnecessary graduate studies.

With this many reasons supporting both courses of action, it is not surprising that there is no consensus. Statistics indicate that roughly 25% of graduating seniors continue directly to graduate programs in electrical engineering and computer engineering and about 55% enter engineering-related industry and government positions after graduating. The remaining 20% pursue a variety of other options including the military, business, law, and medicine. Of students who graduate and enter the engineering industry, about 75% will continue at some point in their career to graduate school, often after first working 3 to 5 years.

Many students seek paid summer internships, often during their rising junior or senior years. Besides earning a substantial amount of money, these often help clarify the choice between industry and graduate school. If the student chooses the former, internships often lead to postgraduation job offers.

COMMAND REVIEW

Function Definition

```
def myFunction(inputParameter1, ...):
    statement(s)
    return ouptutParameter1, ...
```

Function definition with input parameters.
Indented function statement(s).
 return object(s).

Function Calling

```
outputParameter1, ... = myFunction(inputParameters, ...)
```

Module Import

```
import moduleName as nm
nm.moduleFunc()
nm.variable
```

imports file moduleName.py with alias nm.
Use moduleFunc() defined in module nm.
Use variable defined in module nm.

Comments and help()

```
# whole line comment
print('hello') # line comment
```

Entire line ignored by interpreter.
Everything after # ignored by interpreter.

```
def funcDocString():
    """
    docstring comments for help()
    """
    statement(s)
    return object(s)

help(funcDocString)
```

Function with docstring defined.

Define function docstring between " " ".

Body of function.
 return object(s) to calling program.

Show funcDocString() docstring.

Utility Functions

```
import numpy as np
npArray.sort()
npArray.sort(reverse='True')
newArray = np.sort(oldArray)
newDesc = np.sort(oldArray, reverse='True')
sum(v)
```

Needed for NumPy arrays and np.sort().
Ascending sort NumPy array npArray.
Descending sort NumPy array npArray.
Return ascending sorted newArray.
Return descending sorted newDesc.
Returns sum of v values.

Relational statements

```
<, <=
>, >=
==
```

Less than, less than or equal to.
Greater than, greater than or equal to.
Equal to, works with both arrays and strings.

Logical Operators

```
a and b
a or b
```

AND, True if both a and b are True.
OR, True if either a or b is True.

`not a`	NOT, negates the logical value of `a`.
`all(v)`	`True` if all vector `array v` values are `True`.
`any(v)`	`True` if any vector `array v` values are `True`.

Conditional Statements

```
if test:
    print('test is True')
```
if `test` is `True` then
execute this indented code block.

```
if test:
    print('test is True')
else:
    print('test is False')
```
if `test` is `True` then
execute this indented code block.
`else` test is `False` then
execute this indented code block.

`if-elif-else`   Multiple tests with different code blocks.

String Formatting

```
Num = 3.14159
```
str() & +
`'Pi=' + str(Num)`   Creates the string `'Pi=3.14159'`.
f-string or string interpolation
`f'Pi={Num:6.2f}'`   Creates the string `'Pi=  3.14'`.
`f'Pi={Num:<6.2f}'`   Creates the string `'Pi=3.14  '`.
`f'Pi={Num:^6.2f}'`   Creates the string `'Pi= 3.14 '`.
`f'Pi={Num:>6.2f}'`   Creates the string `'Pi=  3.14'`.
.format()
`'Pi={:6.2f}'.format(Num)`   Creates the string `'Pi=  3.14'`.
% operator
`'Pi=%6.2f' %(Num)`   Creates the string `'Pi=  3.14'`.

LAB PROBLEMS

©=Write only the Python command(s).

®=Write only the Python result.

Solutions to all starred (*) problems are available on the Elsevier website (refer to page xix for the link).

4.1* Create a function called `labProblem4_1()` that takes as input the values of two resistors, R1 and R2, and a character code that can be either "S" or "P". Have the function return the equivalent value of the resistors in series if the code is "S" and in parallel if the code is "P". The function definition and `return` statement should be of the form

```
def labProblem4_1(R1, R2, code):
    # statement(s)
    return Rvalue
```

Demonstrate that your function works for both series and parallel cases using the values `R1=10` and `R2=15`. ©®

4.2 Create a function called `labProblem4_2()` that takes two real numbers, a magnitude and an angle in degrees, and returns the corresponding complex number in rectangular form, $z = x + yj$. You may need to review the information about complex numbers from Chapter 2. The function definition and `return` statement should be of the form

```
def labProblem4_2(mag, degrees):
    # statement(s)
    return z
```

To test, `labProblem4_2(10, 30)` should return an answer close to `8.6603 + 5.000j`. Be careful not to confuse radians with degrees. You may have to convert to radians by multiplying your degrees by $\pi/180$. ©®

4.3 Create a function that does the inverse of lab problem 4.2. It should take a single complex number `z` and return its magnitude and angle in degrees. The function definition and `return` statement should be of the form

```
def labProblem4_3(z):
    # statement(s)
    return mag, degrees
```

To test, `labProblem4_3(1+1j)` should return `mag = 1.414` and `degrees = 45`. ©®

4.4* A small college stadium uses 250 high-efficiency LED lighting modules. Each module consumes 500 W of power. If electricity costs $0.000135/W every hour, write a function that takes the number of hours that the lights must be on and returns a string containing the cost of electricity use, in dollars. The function definition and `return` statement should be of the form

```
def labProblem4_4(hours):
    # statement(s)
    return costStr
```

The answer should be rounded to the nearest cent and the string should begin with a "$". For example, `labProblem4_4(7)` should return `$118.13`. ©®

4.5 Voltage dividers, introduced in chapter 2, take a voltage V_1 and reduce it to a voltage V_2 using resistors R_1 and R_2 as shown in Fig. 4.55. This circuit uses a constant current I.

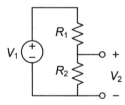

Fig. 4.55 Voltage divider circuit (Lab Problem 4.5).

The two formulas describing this circuit are

$$V_2 = V_1 \frac{R_2}{R_1 + R_2} \text{ and } I = \frac{V_1}{R_1 + R_2}$$ (4.3)

Write a program that takes as arguments V1, V2, and I and returns the resistors R1 and R2 needed to build the circuit. The function definition and `return` statement should be of the form

```
def labProblem4_5(V1, V2, I):
    # statement(s)
    return R1, R2
```

Testing using the call statement `labProblem4_5(12, 4, 0.01)` should return `R1 = 800` and `R2 = 400`. ©®

4.6 A measure of energy used by electrical engineers is the watt-hour, abbreviated W·hr, which is equal to 1 W for 1 h. Solar cell panels provide an annual energy output of 20 W·hrs per square foot of solar cells. Write a function that takes the width and length of a roof surface and returns a string with the expected annual energy output if the roof is covered with solar cells. If the number is less than 1000 W·hrs, return the answer in W·hrs: for example, "936 W·hrs." If the answer is equal to or greater than 1000 W·hrs, such as 2500 W·hrs, convert it into kW·hrs, such as "2.5 kW·hr." The function definition and `return` statement should be of the form

```
def labProblem4_6(width, length):
    # statement(s)
    return energy
```

Testing using the call statement `labProblem4_6(20, 40)` should return `16 kW-hrs`. ©®

4.7* A particular class has an overall semester grade that is entirely dependent on three equally weighted tests. Write a function that takes the first two test grades and a desired overall class grade and returns what the last grade must be to earn the given desired overall grade. The function definition and `return` statement should be of the form

```
def labProblem4_7(g1, g2, gdesired):
    # statement(s)
    return g3
```

Testing using the call statement `labProblem4_7(85, 92, 90)` should return `93`. ©®

4.8 The circuit shown in Fig. 4.56 slowly blinks the LED. It uses two new components called bipolar junction transistors (BJTs). The time between blinks is given by the equation

$$t_{blink} = (0.014)R \tag{4.4}$$

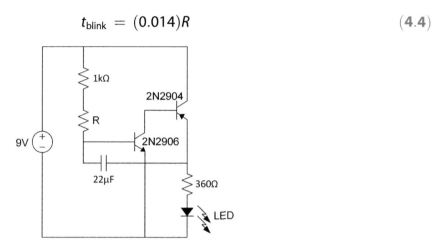

Fig. 4.56 Blinking circuit using BJT (Lab Problem 4.8).

where t_{blink} is measured in seconds. The value of R is measured in kΩ and can range between 10 and 100 kΩ. For example, if R is 22 kΩ, then $t_{blink} = 0.308$ s. Write a program called `labProblem4_8()` that does the opposite. That is, it takes t_{blink} as an input argument and returns the required resistor value R in kΩ. If the value of R is greater than the allowable range of 10–100 kΩ, it instead returns -1. The statement `labProblem4_8(7)` should return `R=-1`. ©®

4.9 A more complicated oscillator circuit than the one described in lab problem 4.8 can create sinusoidal and square waves from 0.01 Hz to 1 MHz. It uses the XR-2200 integrated circuit (IC), as shown in Fig. 4.57.

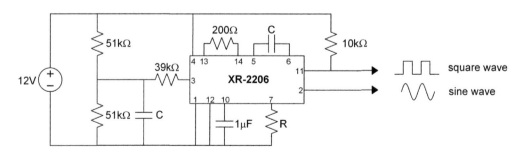

Fig. 4.57 Blinking circuit using XR-2206 (Lab Problem 4.9).

The R and C values control the frequency at which the circuit oscillates according to the equation $f=1/(RC)$, where R can be in the range 4–200 kΩ and C can be in the range 10^{-9}–10^{-4} F. Write a function that takes the value of R in ohms and C in farads and returns either the frequency at which the circuit oscillates if R and C are in range or -1 if they are not in the allowable range. The function definition should be of the form

```
def labProblem4_9(R,C):
    # statement(s)
    return f
```

As an example, `f = labProblem4_9(5e3, 4.7e-5)` sets $f = 4.2553$, indicating oscillation at about 4.26 Hz, but entering `f = labProblem4_9(5e3, 4.7e-4)` sets $f = -1$, since the capacitor value is not in range. ©®

4.10* If you have two resistors, you can use them individually or in combination to get four different resistance values. Write a program that, given two resistors R1 and R2, outputs a vector array with all possible values of resistance in sorted order. That is, R1, R2, and their value in series, and their value in parallel, are sorted from smallest to largest. The function definition is

```
def labProblem4_10(R1, R2):
    # statement(s)
    return Rlist
```

Testing it with `labProblem4_10(10, 15)` should return `[6. 10. 15. 25.]`. ©®

4.11 More challenging: Do the same problem as above but using three resistors. There are many ways to combine them. For instance: R2 alone, R1 in series with R2, R1 in series with R2, and R3 in parallel. This is a tough problem! See how many unique ways you can combine them — the more unique combinations, the higher your score. Sort the resulting values from lowest to highest. The function definition should be

```
def labProblem4_11(R1, R2, R3):
    # statement(s)
    return Rlist
```

Testing it with `labProblem4_11(30, 70, 420)` should return

```
[20. 21. 28. 28.26923077 30. 60. 60.57692308 70. 80.76923077 90. 98. 100.
420. 441. 450. 490. 520.]. ©®
```

4.12 High-pass filters are used to block low frequencies and pass high frequencies. An example of one such filter is shown in Fig. 4.58.

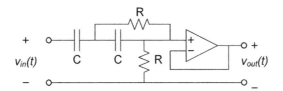

Fig. 4.58 High-pass filter circuit (Lab Problem 4.12).

It has a corner frequency, which is the demarcation between its blocking and passing frequencies, of

$$f = \frac{1}{2\pi RC} \tag{4.5}$$

where *f* is in hertz and *R* and *C* are in ohms and farads, respectively. Create a function called `labProblem4_12()` that takes the values of *R* and *C* and returns the corner frequency *f*. The function definition is

```
def labProblem4_12(R, C):
    # statement(s)
    return f
```
©

4.13* Microcontrollers frequently use 8-bit digital values, which are integers between 0 and 255, inclusive. Create a function that takes a vector of input values and returns its closest integers between 0 and 255. The function definition should be

```
def labProblem4_13(vin):
    # statement(s)
    return result
```

Testing it with `labProblem4_13([-4.2, 0, 0.3, 2.4, 25, 254.5, 255, 999])` should return `[0, 0, 0, 2, 25, 255, 255, 255]`. ©®

4.14 Students decide to use Python to roll a virtual die in a Dungeons & Dragons™ game. Use the `NumPy random.rand()` function with the `NumPy ceil()` function, or the new function `NumPy random.randint()`, with the `sum()` function, that takes an array and returns the sum of its components. The function should take two arguments, the number of rolls and the virtual die's number of sides, and it should return the sum of all the rolls. For example, a single roll of a twenty-sided die would be made using `labProblem4_14(1,20)`, and the sum of three rolls of a six-sided die, which would be some integer between 3 and 18, is found using `labProblem4_14(3,6)`. ©®

4.15 Write a function that improves `engform()` in Fig. 4.52 to work with all the given prefixes. That is, it should take an argument such as 0.0001235 and return "123.5u". Use "u" for "μ". The function definition and `return` statement should be of the form

```
def labProblem4_15(x):
    # statement(s)
    return s
```

Hint: Build it using the starter code in the Tech Tip: Engineering Notation in section 4.15.2. ©®

4.16* The digital circuit shown in Fig. 4.59 has the truth table shown in Table 4.7. This means, for example, that when `x=False`, `y=False`, and `z=True`, `out` will be `False`, but when `x=False`, `y=True`, and `z=False`, `out` will be `True`. Inputs and outputs will always be `True` or `False`.

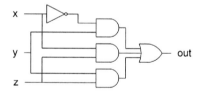

Fig. 4.59 Digital logic circuit (Lab Problem 4.16).

Create a function called `labPoblem4_16()` that takes the values of `x`, `y`, and `z` and returns the result `out`. The function definition should be of the form

```
def labProblem4_16(x, y, z):
    # statement(s)
    return out
```

TABLE 4.7 Digital logic circuit truth table

x	y	z	out
False	False	False	False
False	False	True	False
False	True	False	True
False	True	True	True
True	False	False	False
True	False	True	True
True	True	False	False
True	True	True	True

For example, `labProblem4_16(False,True,False)` should return `True`. Use the truth table to check your work. ©®

4.17 Challenging! Write a function called `labProblem4_17()` that takes a resistance value that is an integer from 1 to 1,000,000 and outputs a string of four colors that correspond to the resistor's color code for a 5% tolerance resistor. Do not capitalize the colors, but put spaces between them. The function definition should be of the form

```
def labProblem4_17(x):
    # statement(s)
    return s
```

Testing it by calling the function using `labProblem4_17(10000)` should return the Python string

```
'brown black orange gold'
```

Hint: Use the `if-elif` statement to print the color strings. ©®

PROGRAMMING II: LOOPING

5.1 OBJECTIVES

After completing this chapter, you will be able to use Python to do the following:

- Program using `for` loops
- Write `for` loops to iterate over a `list` of objects
- Write `for` loops using the `range()`, NumPy `arange()`, and `enumerate()` functions
- Include `else`, `break`, and `continue` statements in `for` loops
- Nest loops inside each other
- Program using `while` loops
- Write `while` loops using counters, `else` blocks, `break`, and `continue` statements
- Preallocate `array` memory to speed up program execution
- Perform basic binary logic commands

5.2 LOOPING IN PYTHON PROGRAMS

Looping is a powerful programming technique that allows Python users to execute a set of code statements for each object in a collection or while a logical statement tests `True`. For example, you may need to run a simulation in which a filter's response to variations based on a resistor's tolerance is calculated and displayed for each of 1000 random perturbations in resistance. You may also need to keep adding successive terms in a numerical series until the error decreases below an appropriate value. The `for` and `while` loops can be used for such applications.

5.3 FOR LOOP

A `for` loop can be used to iterate over a collection of objects. The basic `for` loop code structure is outlined in Fig. 5.1, and the corresponding flowchart is shown in Fig. 5.2. During each iteration, an object of the iterable collection is assigned to the variable `var` and one or more statements in the indented body of the `for` loop are executed. After the last statement in the `for` loop body has been executed, the loop begins again by assigning the next object of the collection to the variable `var`. After all the objects of the collection have been assigned to the variable `var`, the `for` loop exits and program flow continues to the line following the indented `for` loop statement(s).

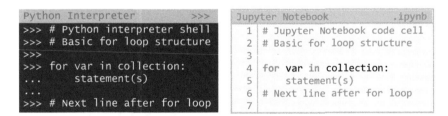

Fig. 5.1 Basic `for` loop code structure. An object `in` the collection is assigned to the variable `var`, and the indented body of the `for` loop is entered. Upon completion of the `for` loop statement(s), the next object `in` the collection is assigned to `var` and the loop repeats. After all objects `in` the collection have been assigned to `var`, the loop exits and program control moves to the first line following the indented `for` loop statement(s).

5.3.1 For Loop Iteration over a List of Objects

One of the most important collections of objects that can be used in a `for` loop is the `list`. A `list` is a sequential collection of comma-separated objects enclosed by square brackets `[]`.

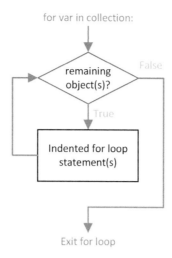

for var in collection:

remaining object(s)?

False

True

Indented for loop statement(s)

Exit for loop

Fig. 5.2 Basic `for` loop flowchart. The `for` loop iterates over each object `in` a collection. After every object has been iterated through, the `for` loop exits and program control moves to the line following the indented `for` loop statement(s).

The objects can be any data type, such as strings (`str`), numbers (`int`, `float`, & `complex`), or Booleans (`bool`). For example, you could create a `list` of courses in your first semester:

```
>>> courses = ["Python", "Calculus", "Physics", "Chemistry"]
```

Recall that 'single' or "double" quotation marks define a string. You can iterate over your courses using a `for` loop and `print` out each object `in` your courses `list` as demonstrated in Fig. 5.3.

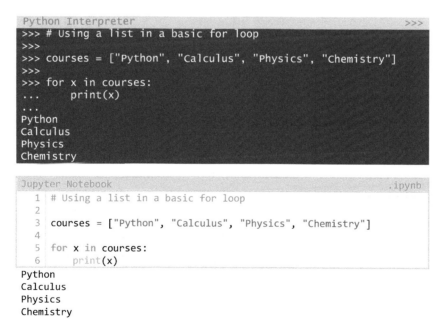

```
Python Interpreter                                          >>>
>>> # Using a list in a basic for loop
>>>
>>> courses = ["Python", "Calculus", "Physics", "Chemistry"]
>>>
>>> for x in courses:
...     print(x)
...
Python
Calculus
Physics
Chemistry
```

```
Jupyter Notebook                                          .ipynb
1  # Using a list in a basic for loop
2
3  courses = ["Python", "Calculus", "Physics", "Chemistry"]
4
5  for x in courses:
6      print(x)
Python
Calculus
Physics
Chemistry
```

Fig. 5.3 Using a `list` in a `for` loop. A `list` is a collection of comma separated objects enclosed by square brackets. During each iteration of the `for` loop, a `list` object is assigned to the `for` variable `x`.

If your `list` has many objects, you may wish to use the backslash character, `\`, to continue your code to the next line. For example, you could have written your `list` of courses as

```
>>> courses = ["Python", "Calculus", \
...            "Physics", "Chemistry"]
```

or as

```
>>> courses = ["Python", \
...            "Calculus", \
...            "Physics", \
...            "Chemistry"]
```

5.3.2 For Loop Iteration Using the Built-in range() Function

Python has a special built-in function, `range()`, and a `NumPy arange()` function that are commonly used in `for` loops. The function `range()` creates a sequence of `int`, or integer, objects using the format

```
>>> range(start, stop, step)
>>> range(start, stop)        # Default step = 1
>>> range(stop)               # Default start = 0 and default step = 1
```

This produces `int` objects beginning at `start` and incrementing by `step` up to, but not including, the `stop` value. Figure 5.4 illustrates all three of these cases using examples.

5.3.3 For Loop Iteration Using the NumPy arange() Function

The `NumPy arange()` function is a more powerful alternative to the built-in `range()` function. The `NumPy` function `arange()` returns a sequential `NumPy array` using the same format as the `range()` function. The `NumPy arange()` function is faster and can generate a sequential `array` of `float`, or real, numbers.

```
>>> import numpy as np
>>> np.arange(start, stop, step)
>>> np.arange(start, stop)            # Default step = 1
>>> np.arange(stop)                   # Default start = 0 and default step = 1
```

(a)

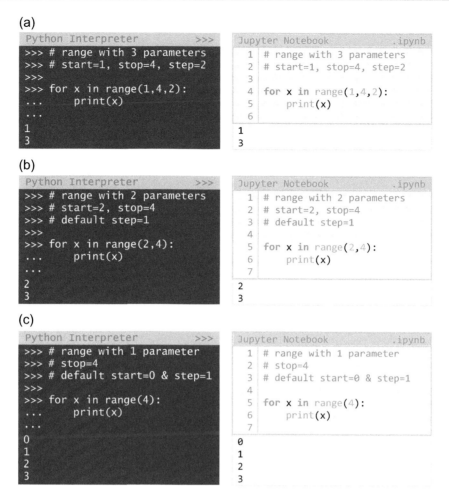

(b)

(c)

Fig. 5.4 Using the built-in `range()` function in a `for` loop. (a) If all three arguments of the `range()` function are specified, the first is the `start` value, the second the `stop` value, and the third the `step` value. (b) If only two input arguments are specified, the first is the `start` value and the second the `stop` value. The `step` value defaults to one in this case. (c) If only one input argument is specified, it is assumed to be the `stop` value. In this case the `start` value defaults to zero and the `step` value defaults to one.

Fig. 5.5 demonstrates the use of the `NumPy arange()` function in a `for` loop to iterate over a sequence of `float`, or real number, objects.

The iteration variable `x` can also be used in numerical operations within the body of the `for` loop. For example, the code shown in Fig. 5.6 prints the squares and cubes of three even numbers.

5.3.4 For Loop Iteration Using the General NumPy array Collections

One of the most common uses of the `for` loop is to iterate through numerical objects. Using the built-in `range()` or `NumPy arange()` functions to produce evenly spaced arrays of numerical objects is one example of this. A `for` loop, however, can iterate over a more

```
Python Interpreter                                          >>>
>>> # Using the numpy arange() function in a for loop
>>> # numpy arange() generates a linear sequence of real numbers
>>> # start=1.5, stop=3.0, step=0.5
>>>
>>> import numpy as np
>>>
>>> v = np.arange(1.5, 3.0, 0.5)
>>> print("v =", v)
v = [1.5 2.  2.5]
>>>
>>> for x in v:
...     print("x =", x)
...
x = 1.5
x = 2.0
x = 2.5
```

```
Jupyter Notebook                                         .ipynb
 1  # Using the numpy arange() function in a for loop
 2  # numpy arange() generates a linear sequence of real numbers
 3  # start=1.5, stop=3.0, step=0.5
 4
 5  import numpy as np
 6
 7  v = np.arange(1.5, 3.0, 0.5)
 8  print("v =", v)
 9
10  for x in v:
11      print("x =", x)
v = [1.5 2.  2.5]
x = 1.5
x = 2.0
x = 2.5
```

Fig. 5.5 Using the NumPy arange() function in a for loop. The NumPy arange() function creates a linearly spaced array of real numbers that the for loop iterates over. During each iteration of the for loop, a value of the array v is assigned to the variable x. Unlike the range() function, which can only generate int, or integer, sequences, the NumPy arange() function produces float, or real number, sequential arrays.

```
Python Interpreter            >>>
>>> # Print squares & cubes
>>>
>>> import numpy as np
>>>
>>> for x in np.arange(2,7,2):
...     print(x, x**2, x**3)
...
2 4 8
4 16 64
6 36 216
```

```
Jupyter Notebook        .ipynb
1  # Print squares & cubes
2
3  import numpy as np
4
5  for x in np.arange(2,7,2):
6      print(x, x**2, x**3)
7
2 4 8
4 16 64
6 36 216
```

Fig. 5.6 Performing numerical operations in for loops. A sequence of squares and cubes can be produced using the iteration variable x.

general array of numerical objects that are not necessarily evenly spaced. Consider, for example, a function that prints a set of standard resistor values from an array, as illustrated in Fig. 5.7.

```
Python Interpreter                                              >>>
>>> # Function to print standard resistor values
>>>
>>> import numpy as np
>>>
>>> def printStdResistorValues():
...     stdResistorValues = np.array([10,15,22,33,47,68,82])
...     for stdResistor in stdResistorValues:
...         print(stdResistor)
...     return None
...
>>> printStdResistorValues()
10
15
22
33
47
68
82
```

```
Jupyter Notebook                                              .ipynb
 1  # Function to print standard resistor values
 2
 3  import numpy as np
 4
 5  def printStdResistorValues():
 6      stdResistorValues = np.array([10, 15, 22, 33, 47, 68, 82])
 7      for stdResistor in stdResistorValues:
 8          print(stdResistor)
 9      return None
10
11  printStdResistorValues()
10
15
22
33
47
68
82
```

Fig. 5.7 Using a `for` loop to iterate through an `array` of numerical objects. An `array` of nonuniformly spaced resistor values can be iterated over in a `for` loop and printed out.

PRACTICE PROBLEMS

5.1   What does the function `practiceProblem5_1(n)` in the code listing below do?

```
# Practice problem 5.1
def practiceProblem5_1(n):
    x = 1
    for i in range(n):
        x = x + x
```

```
        print(x)

    return None
```

Try to figure it out before you run it. Check by passing in an `n` value between 5 and 10. For example, executing the statement `practiceProblem5_2(5)` will test the function with an `n` value equal to 5. ®

5.2   Fill in the missing part in the program listing

```
# Practice problem 5.2
def practiceProblem5_2(r):
    for r2 in [10, 15, 22, 33, 47, 68, 82]:
        rp = (*** function of r and r2 ***)
        print(rp)

    return None
```

labeled `***` to create a function that takes a single resistor value `r`, returns nothing, and displays the value of `r` in parallel with the list `[10, 15, 22, 33, 47, 68, 82]` of standard resistor values using a `for` loop. Test with an input resistor value of 30. ©®

5.3   Write a function that takes no arguments, returns no arguments, and prints two columns showing the conversion from degrees to radians. Print from 0 to 180 degrees in 5-degree increments. ©

5.4 FOR LOOPS ITERATING VECTOR ARRAYS

One of the most common uses of a `for` loop is to iterate through the numerical objects of a `NumPy array`. This permits complex operations to be applied to vector `arrays`. For example, a different way to accomplish practice problem 5.2 is with the function listed in Fig. 5.8. The function `forArrayIterationExample()` begins by defining a `NumPy array`, called

```
Python Interpreter                                            >>>
>>> # Parallel resistances formed with a standard resistor set
>>>
>>> import numpy as np
>>>
>>> def forArrayIterationExample(rInput):
...     """
...     Using an input resistor, rInput, this function
...     calculates the parallel resistance, rParallel, with
...     each resistor of a set of standard resistors and
...     prints out the parallel value.
...     Inputs
...         rInput: a single input resistance
...     Returns
...         None
...     """
...     stdResistors = np.array([10, 15, 22, 33, 47, 68, 82])
...     for rObject in stdResistors:
...         rParallel = (rInput*rObject)/(rInput+rObject)
...         print(rParallel)
...     return None
...
>>>
```

```
Jupyter Notebook                                          .ipynb
 1  # Parallel resistances formed with a standard resistor set
 2
 3  import numpy as np
 4
 5  def forArrayIterationExample(rInput):
 6      """
 7      Using an input resistor, rInput, this function
 8      calculates the parallel resistance, rParallel, with
 9      each resistor of a set of standard resistors and
10      prints out the parallel value.
11      Inputs
12          rInput: a single input resistance
13      Returns
14          None
15      """
16      stdResistors = np.array([10, 15, 22, 33, 47, 68, 82])
17      for rObject in stdResistors:
18          rParallel = (rInput*rObject)/(rInput+rObject)
19          print(rParallel)
20      return None
```

Fig. 5.8 Using a for loop to form parallel resistances with a set of standard resistances and a single input resistance. The parallel resistance formed during each iteration is printed out.

stdResistors, that holds a set of standard resistor values. Then it creates a for loop variable rObject to iterate through each array object. During each iteration, a resistor object in the stdResistors array is assigned to rObject. Then it finds the parallel combination of the rObject value with the input resistor value and displays the result.

Up to this point, `array` math has been handled with a single operation. For example, multiplication of `array v` by a constant `c` is the single operation `v*c`. This operates on the entire array at once and is very efficient, but some operations are more easily understood using a `for` loop that steps through every value inside the vector array. For instance, consider a piecewise linear function called the step function and abbreviated $u(t)$, defined as

$$u(t) = \begin{cases} 0, & t < 0 \\ 1, & t \geq 0 \end{cases} \qquad (5.1)$$

and graphed in Fig. 5.9.

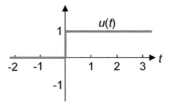

Fig. 5.9 Unit step function, $u(t)$. The unit step function is zero for all t values less than zero and equal to 1 for all t values greater than or equal to zero. This function is often used in systems to simulate turning a switch on at $t = 0$.

A function that generates a plot of the unit step function is given in Fig. 5.10. Notice how the `for` loop sets `i` to every index of the `vt` vector `array` and then individually performs some operation on it. In this application, `vt` is checked to see if it is negative, and then the output vector `array` is created accordingly.

Figure 5.11 lists a program for generating a more complex piecewise plot of the mathematical function

$$f(t) = \begin{cases} \sqrt{10t}, & 0 \leq t < 10 \\ 10e^{10-t}, & 10 \leq t < 20 \\ t - 20, & 20 \leq t \leq 30 \end{cases} \qquad (5.2)$$

```
Python Interpreter                                                    >>>
>>> # Unit step function plot using built-in range() function
>>>
>>> import numpy as np
>>> import matplotlib.pyplot as plt
>>>
>>> def plotStepFunction():
...     """Plots the unit step function"""
...     vt = np.linspace(-2,3,1000)   # Horizontal axis, 1000 pts
...     vu = np.zeros(1000)           # Preallocate function
...     # Use built-in length function, len, to get vt length
...     for i in range(len(vt)):      # Iterate over length of vt
...         if  vt[i] < 0.0:          # If vt[i] is negative
...             vu[i] = 0.0           #     set vu[i] to 0.0
...         else:                     # otherwise
...             vu[i] = 1.0           #     set vu[i] to 1.0
...     fig, ax = plt.subplots()      # Set up plot figure & axis
...     ax.plot(vt,vu)                # Plot vu versus vt
...     ax.set_xlabel("t")            # x axis label
...     ax.set_ylabel("u(t)")         # y axis label
...     plt.show()                    # Show the plot
...     return None                   # return None object
...
>>>
```

```
Jupyter Notebook                                               .ipynb
 1   # Unit step function plot using built-in range() function
 2
 3   import numpy as np
 4   import matplotlib.pyplot as plt
 5   %matplotlib inline
 6
 7   def plotStepFunction():
 8       """Plots the unit step function"""
 9       vt = np.linspace(-2,3,1000)   # Horizontal axis, 1000 pts
10       vu = np.zeros(1000)           # Preallocate function
11       # Use the built-in length function, len, to get vt length
12       for i in range(len(vt)):      # Iterate over length of vt
13           if  vt[i] < 0.0:          # If vt[i] is negative
14               vu[i] = 0.0           #     set vu[i] to 0.0
15           else:                     # otherwise
16               vu[i] = 1.0           #     set vu[i] to 1.0
17       fig, ax = plt.subplots()      # Set up a plot figure & axis
18       ax.plot(vt, vu)               # Plot vu versus vt
19       ax.set_xlabel("t")            # x axis label
20       ax.set_ylabel("u(t)")         # y axis label
21       return None                   # return None object
```

Fig. 5.10 Plotting the unit step function. The range() and len() functions generate the indices for the vector arrays to produce the unit step function.

from $0 \leq t \leq 30$. Figure 5.12 shows the plot produced by this program. Notice how the variable Np holds the number of data points and is used at several different points in the program. Changing the number of data points becomes a simple matter of changing the value at one point. Having to change the numerical value at several different points in a program creates more opportunities to introduce bugs. In addition, numerical values provide little descriptive information regarding what the number is for.

```
Python Interpreter                                           >>>
>>> # Piecewise function plot using numpy.arange() function
>>> # numpy.arange(Np) generates an array from 0 to Np-1
>>>
>>> import numpy as np
>>> import matplotlib.pyplot as plt
>>>
>>> def plotPiecewiseFunction():
...     """
...     Plots a piecewise function
...     Inputs:  None
...     Returns: None
...     """
...
...     # Let t range from 0 to 30 with Np=1000 points
...     Np = 1000 # Variable Np holds the number of data points
...     vt = np.linspace(0.0,30.0,Np) # Time values data array
...     # Preallocate function array and fill with zeros
...     vf = np.zeros(Np) # Create numpy array with Np zeros
...     for i in np.arange(Np): # Iterate indices 0 through Np-1
...         t = vt[i]                # Iteration t value
...         if  (0.0<=t) and (t<10.0):     # if 0<=t<10
...             vf[i] = np.sqrt(10.0*t)  #       vf=sqrt(10t)
...         elif (10.0<=t) and (t<20.0):   # else if 10<=t<20
...             vf[i]=10.0*np.exp(10.0-t) #      vf=10exp(10-t)
...         elif (20.0<=t):                # else if 20<=t
...             vf[i] = t-20.0             #      vf=t-20
...     # Plot the function
...     fig, ax = plt.subplots() # Set up plot figure & axis
...     ax.plot(vt,vf)            # Plot vf versus vt
...     ax.set_xlabel("t")        # x axis label
...     ax.set_ylabel("f(t)")     # y axis label
...     plt.show()    # show() needed in Python interpreter shell
...     return None   # No objects, or None, are returned
...
>>>
```

```
Jupyter Notebook                                          .ipynb
 1  # Piecewise function plot using numpy.arange() function
 2  # numpy.arange(Np) generates an array from 0 to Np-1
 3
 4  import numpy as np
 5  import matplotlib.pyplot as plt
 6  %matplotlib inline
 7
 8  def plotPiecewiseFunction():
 9      """
10      Plots a piecewise function
11      Inputs:  None
12      Returns: None
13      """
14      # Let t range from 0 to 30 with Np=1000 points
15      Np = 1000 # Variable Np holds the number of data points
16      vt = np.linspace(0.0,30.0,Np) # Time values data array
17      # Preallocate function array and fill with zeros
18      vf = np.zeros(Np) # Create numpy array with Np zeros
19      for i in np.arange(Np): # Iterate indices 0 through Np-1
20          t = vt[i]                # Iteration t value
21          if  (0.0<=t) and (t<10.0):     # if 0<=t<10
22              vf[i] = np.sqrt(10.0*t)   #      vf=sqrt(10t)
23          elif (10.0<=t) and (t<20.0):  # else if 10<=t<20
24              vf[i]=10.0*np.exp(10.0-t) #     vf=10exp(10-t)
25          elif (20.0<=t):               # else if 20<=t
26              vf[i] = t-20.0            #     vf=t-20
27      # Plot the function
28      fig, ax = plt.subplots() # Set up plot figure & axis
29      ax.plot(vt,vf)            # Plot vf versus vt
30      ax.set_xlabel("t")        # x axis label
31      ax.set_ylabel("f(t)")     # y axis label
32      return None   # No objects, or None, are returned
```

Fig. 5.11 Plotting piecewise functions using a `for` loop and the `NumPy arange()` function. An `if-elif-elif` structure is used to define the different domains of the function.

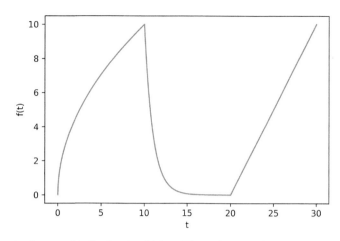

Fig. 5.12 Piecewise function, $f(t)$. The function $f(t)$ has different functional forms over different time intervals.

PRACTICE PROBLEMS

5.4 Create a function using the `numpy arange()` function that plots the function

$$v(t) = \begin{cases} 1 - t, & t < 0 \\ e^{t/4}, & t \geq 0 \end{cases} \tag{5.3}$$

for $-5 \leq t \leq 5$. Plot using 100 points and label each axis. ®

In the previous two examples the `range()` and `NumPy arange()` functions were used to generate iteration indices for the vector arrays. Using the `enumerate()` function it is possible to extract both the index and object from an iterable collection. Using the `enumerate()` function, when possible, is generally considered good Python practice. The solution to Practice Problem 5.4 using the `enumerate()` function is illustrated in Fig. 5.13.

```
Python Interpreter                                               >>>
>>> # Piecewise function plot using the enumerate() function
>>> # enumerate(collection) returns both the current index
>>> # and object of the iterable collection during a for loop
>>>
>>> import numpy as np
>>> import matplotlib.pyplot as plt
>>>
>>> def plotPP5_4Using_enumerate():
>>>     """
>>>     Plots a piecewise linear and exponential function
...         using enumerate()
...     Input parameters: None
...     Returns: None
...     """
...     # Let t range from -5.0 to 5.0 using Np=100 points
...     Np = 100  # Variable Np holds the number of data points
...     vt = np.linspace(-5.0,5.0,Np)    # Time values data array
...     # Preallocate function vector array and fill with zeros
...     vf = np.zeros(Np) # Create array filled with Np zeros
...     for index,t in enumerate(vt):    # Get index and t value
...         if (t<0.0):                  # If t<0
...             vf[index] = 1.0-t        #       vf = 1.0-t
...         else:                        # Otherwise
...             vf[index]=np.exp(t/4.0)  #       vf = exp(t/4.0)
...     # Plot the function
...     fig, ax = plt.subplots()        # Set up plot figure & axis
...     ax.plot(vt,vf)                  # Plot vf versus vt
...     ax.set_xlabel("t")              # x axis label
...     ax.set_ylabel("f(t)")           # y axis label
...     plt.show()   # show() needed in Python interpreter shell
...     return None  # No objects, or None, are returned
...
>>>
```

```
Jupyter Notebook                                               .ipynb
 1  # Piecewise function plot using the enumerate() function
 2  # enumerate(collection) returns both the current index
 3  # and object of the iterable collection during a for loop
 4
 5  import numpy as np
 6  import matplotlib.pyplot as plt
 7  %matplotlib inline
 8
 9  def plotPP5_4Using_enumerate():
10      """
11      Plots a piecewise linear and exponential function
12          using enumerate()
13      Input parameters: None
14      Returns: None
15      """
16      # Let t range from -5.0 to 5.0 using Np=100 points
17      Np = 100  # Variable Np holds the number of data points
18      vt = np.linspace(-5.0,5.0,Np)    # Time values data array
19      # Preallocate function vector array and fill with zeros
20      vf = np.zeros(Np) # Create array filled with Np zeros
21      for index,t in enumerate(vt):    # Get index and t value
22          if (t<0.0):                  # If t<0
23              vf[index] = 1.0-t        #       vf = 1.0-t
24          else:                        # Otherwise
25              vf[index]=np.exp(t/4.0)  #       vf = exp(t/4.0)
26      # Plot the function
27      fig, ax = plt.subplots()        # Set up plot figure & axis
28      ax.plot(vt,vf)                  # Plot vf versus vt
29      ax.set_xlabel("t")              # x axis label
30      ax.set_ylabel("f(t)")           # y axis label
31      return None  # No objects, or None, are returned
```

Fig. 5.13 Solution to Practice Problem 5.4 using the `enumerate()` function. Notice that `enumerate()` returns both the current `index` and the array object `t` in a given iteration.

Monte Carlo simulations are commonly used in EE to test how engineering designs respond to random changes in component values. For example, consider the low-pass circuit shown in Fig. 5.14. This circuit passes signals with frequencies below a cutoff frequency but blocks signals with frequencies above the cutoff frequency. The cutoff frequency in Hz, f_c, is given by

$$f_c = \frac{1}{2\pi RC} \tag{5.4}$$

Fig. 5.14 Low-pass filter. The resistor, R, and capacitor, C, set the cutoff frequency, f_c. Signals with frequencies above the cutoff frequency are attenuated, while those below the cutoff frequency are passed relatively unattenuated.

where R is the resistor value in ohms and C the capacitor value in farads.

Consider choosing $R = 1$ kΩ and $C = 1$ μF to make the filter pass frequencies below 160 Hz. How will this cutoff frequency vary, given a resistor tolerance of $\pm$ 5% and a capacitor tolerance of $\pm$ 10%? A graduate-level course in stochastic theory will derive the result exactly, but it is simpler to simulate the result several thousand times, each time with different random resistor and capacitor values.

In Python, the `NumPy random.rand()` function returns a random number `n` over the interval $0 \leq n < 1$, that is, [0, 1). To change this interval to [−0.05, +0.05) to simulate a 5% resistor tolerance, start with the desired span of 0.1 associated with the interval [−0.05, 0.05). Use `random.rand()*0.1` to get the correct interval [0, 0.1). To center this around zero to give the correct interval [−0.05, 0.05) subtract 0.05, that is, `random.rand()*0.1−0.05`. Multiplying this factor by the given resistor value, $R = 1000$, will scale the random number to get the appropriate error, `1000*(random.rand()*0.1 −0.05)`. This is the random value that added to the original resistor value, $R = 1000$, will create a random resistor value with the specified tolerance of 5%. Similarly, using `NumPy random.rand*0.2−0.1` to produce random numbers from −0.1 to 0.1 for the capacitors will allow you to create a set of random capacitor values around a mean value with specified tolerance. See Fig. 5.15.

```
#Tech Tip: Monte Carlo Simulation Program

import numpy as np

def MonteCarlo(N):
    """
    Runs a total of N Monte Carlo simulations using mean
    values R = 1000 Ohms and C = 1e-6 Farads to calculate the
    cutoff frequency using fc = 1/(2piRC)
    Input parameters
        N: The number of Monte Carlo simulations
    Returns
        fcArray: NumPy array of N cutoff frequencies
    """
    fcArray = np.zeros(N)      # Preallocate with N zeros
    for i in range(N):          # Do simulation N times
        # Use +- 5% resistor tolerance
        R = 1000.0 + 1000.0 * (np.random.rand() * 0.1 - 0.05)
        # Use +- 10% capacitor tolerance
        C = 1e-6 + 1e-6 * (np.random.rand() * 0.2 - 0.1)
        # Calculate the cutoff frequency
        fc = 1.0 / (2.0 * np.pi * R * C)
        fcArray[i] = fc # Store in array
    return fcArray # Return NumPy array with N fc simulations
```

Fig. 5.15 Monte Carlo simulation program listing. During each iteration, random noise is added to a resistor with a mean value of 1000 and a capacitor with a mean value of 1e−6. The cutoff frequency is calculated and stored in an `array`. After all iterations have been completed, the `array` of all cutoff frequencies is returned to the calling program.

To produce a scatterplot, run the function shown in Fig. 5.16 with an input parameter of 1000.

```
#Tech Tip: Run Monte Carlo Simulation and show scatter plot

import matplotlib.pyplot as plt
# Only use %matplotlib inline with a Jupyter notebook
%matplotlib inline

result = MonteCarlo(1000)

fig, ax = plt.subplots()
ax.plot(result, '.')
ax.set_xlabel('Simulation Number')
ax.set_ylabel('Cutoff Frequency, $f_c$ (Hz)')
#plt.show()   # Use plt.show() when using a Python interpreter
```

Fig. 5.16 Driver program for running the Monte Carlo simulation and plotting the results as a scatterplot.

To produce a histogram bar chart, the same Monte Carlo program can be used, but the results can be formatted in a bar chart as shown in Figs. 5.18 and 5.19.

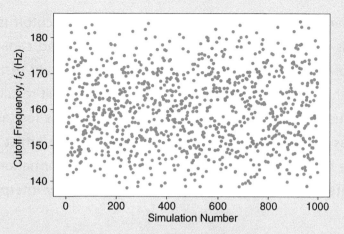

Fig. 5.17 Scatterplot of Monte Carlo cutoff frequency simulations. Adding random noise to the resistor and capacitor values based on their tolerances produces cutoff frequencies that are also randomly distributed.

```
#Tech Tip: Run Monte Carlo simulation and show histogram plot

import matplotlib.pyplot as plt
# Only use %matplotlib inline with a Jupyter notebook
%matplotlib inline

result = MonteCarlo(1000)

fig, ax = plt.subplots()
ax.hist(result, edgecolor = 'black')
ax.set_xlabel('Cutoff Frequency, $f_c$ (Hz)')
ax.set_ylabel('Number'); # Suppress the output
#plt.show()    # Use plt.show() when using a Python interpreter
```

Fig. 5.18 Code to run a Monte Carlo simulation and plot the results as a histogram bar chart.

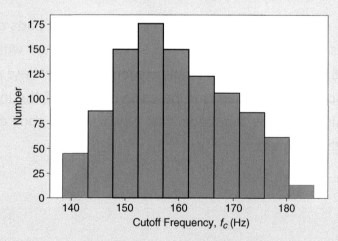

Fig. 5.19 Histogram bar chart of Monte Carlo cutoff frequency simulations. The height of each rectangle shows the number of simulations that produced cutoff frequencies with the interval shown by the horizontal width of the rectangle.

Now it is apparent that the average value of the filters' cutoff is about 160 Hz. This is what would be obtained with resistors and capacitors whose values were precisely as marked. However, using real-world resistors and capacitors with a 5% and 10% tolerance, respectively, will result in a spread of cutoff frequencies. One could expect to find cutoffs as low as 140 Hz and as high as 180 Hz when using components with imperfect real-world values. Notice how the histogram displays the same information as the previous Monte Carlo function does, but in a more easily interpreted format.

PRACTICE PROBLEMS

5.5 Solve the Monte Carlo filter problem, but now assume that the capacitors are higher-quality 5% versions and the resistors have a 1% tolerance. Plot the results of 1000 simulations as a scatterplot and histogram. ®

5.5 NESTED LOOPS

It is possible to nest `for` loops inside each other. For example, if a parts cabinet were stocked with exactly four types of resistors, 10, 22, 47, and 51 Ω, the program with a set of nested `for` loops outlined in Fig. 5.20 would find all possible combinations for these resistor values. Study the output carefully so that you understand how the iteration of the inner and outer loops progress. If you are using a Jupyter Notebook, note how a second code cell is used to call the code cell containing the program.

5.6   Modify the program listed in Fig. 5.20 to print the equivalent resistance of all possible values of R_1 and R_2 in parallel. ©®

5.7   Is it possible to sort the results from practice problem 5.6 with no code changes other than by simply inserting the `sort()` command in the correct place? If so, do it and report the results. If not, explain why it is not possible. This problem should clarify when it is helpful to print results to the screen and when it is helpful to return results as a function's output argument.

```
Python Interpreter                                       >>>
>>> # Nested for loop
>>>
>>> import numpy as np
>>>
>>> def findResistors():
...     """
...     Finds all combinations of 2 resistors given
...         4 stocked resistor types
...     Input parameters: None
...     Returns: None
...     """
...     # Store 4 types of stocked resistors in a numpy array
...     stockedResistors = np.array([10, 22, 47, 51])
...     # Assign 1st resistor to R1
...     for R1 in stockedResistors:
...         # Assign 2nd resistor to R2
...         for R2 in stockedResistors:
...             # Use f-string to format the output
...             print(f"R1 = {R1:2d}, R2 = {R2:2d}")
...     return None # Return no objects, i.e., None
...
>>> findResistors()
R1 = 10, R2 = 10
R1 = 10, R2 = 22
R1 = 10, R2 = 47
R1 = 10, R2 = 51
R1 = 22, R2 = 10
R1 = 22, R2 = 22
R1 = 22, R2 = 47
R1 = 22, R2 = 51
R1 = 47, R2 = 10
R1 = 47, R2 = 22
R1 = 47, R2 = 47
R1 = 47, R2 = 51
R1 = 51, R2 = 10
R1 = 51, R2 = 22
R1 = 51, R2 = 47
R1 = 51, R2 = 51
```

Fig. 5.20 Using nested `for` loops to find all combinations of two resistors from a stock of four different resistor values. Note that the iteration value of the outer loop stays constant until the inner loop finishes iterating through all its values.

```
Jupyter Notebook                                                    .ipynb
  1  # First code cell with nested for loop function definition
  2
  3  import numpy as np
  4
  5  def findResistors():
  6      """
  7      Finds all combinations of 2 resistors given
  8        4 stocked resistor types
  9      Input parameters: None
 10      Returns: None
 11      """
 12      # Store 4 types of stocked resistors in a numpy array
 13      stockedResistors = np.array([10, 22, 47, 51])
 14      # Assign 1st resistor to R1
 15      for R1 in stockedResistors:
 16          # Assign 2nd resistor to R2
 17          for R2 in stockedResistors:
 18              # Use f-string to format the output
 19              print(f"R1 = {R1:2d}, R2 = {R2:2d}")
 20      return None # Return no objects, i.e., None
```

```
  1  # Second code cell that calls findResistors() function
  2
  3  findResistors()
```

```
R1 = 10, R2 = 10
R1 = 10, R2 = 22
R1 = 10, R2 = 47
R1 = 10, R2 = 51
R1 = 22, R2 = 10
R1 = 22, R2 = 22
R1 = 22, R2 = 47
R1 = 22, R2 = 51
R1 = 47, R2 = 10
R1 = 47, R2 = 22
R1 = 47, R2 = 47
R1 = 47, R2 = 51
R1 = 51, R2 = 10
R1 = 51, R2 = 22
R1 = 51, R2 = 47
R1 = 51, R2 = 51
```

Fig. 5.20, cont'd

5.5.1 *Using Nested Loops to Search for* Exact *Solutions*

Consider a program that searches for integer solutions for the sides of a right triangle, known as Pythagorean triples, satisfying $c = \sqrt{a^2 + b^2}$. The program searches for integer hypotenuses as side a ranges from 1 to some given upper bound N and as side b ranges from 1 to N.

RECALL

x is an integer only if x == np.floor(x).

Every time an integer hypotenuse c is found, it is printed to the command window, as shown in Fig. 5.21.

```
Python Interpreter                                              >>>
>>> # Pythagorean search for integer lengths of right triangles
>>>
>>> import numpy as np
>>>
>>> def pythagorean(N):
...     """
...         Search for integer sides of right triangles
...             with side lengths varying from 1 to N
...         Input parameters
...             N: Maximum side length.  Sides vary from 1 to N.
...         Returns: None
...     """
...     # Careful! range(1,N) generates (1, ..., N-1)
...     # range(1,N+1) start=1, stop=N+1 generates (1, ..., N)
...     for a in range(1,N+1):          # a values from 1 to N
...         for b in range(1,N+1):   # b values from 1 to N
...             c = np.sqrt(a*a+b*b) # Find hypotenuse c
...             if (c==np.floor(c)): # if c is an integer
...                 print(c)          #    print out the value
...     return None
...
>>>
```

```
Jupyter Notebook                                           .ipynb
 1 | # Pythagorean search for integer lengths of right triangles
 2 |
 3 | import numpy as np
 4 |
 5 | def pythagorean(N):
 6 |     """
 7 |     Search for integer sides of right triangles
 8 |       with side lengths varying from 1 to N
 9 |     Input parameters
10 |         N: Maximum side length. Sides vary from 1 to N.
11 |     Returns: None
12 |     """
13 |     # Careful! range(1,N) generates (1, ..., N-1)
14 |     # range(1,N+1) start=1, stop=N+1 generates (1, ..., N)
15 |     for a in range(1,N+1):        # a values from 1 to N
16 |         for b in range(1,N+1):   # b values from 1 to N
17 |             c = np.sqrt(a*a+b*b) # Find hypotenuse c
18 |             if (c==np.floor(c)): # If c is an integer
19 |                 print(c)          #    print out the value
20 |     return None
```

Fig. 5.21 Finding the integer solutions to right triangles. This program finds the integer solutions to right triangles having side values ranging from 1 to a given input value *N*. The algorithm is implemented using two nested `for` loops and an `if` statement to test if the hypotenuse is an integer value.

PRACTICE PROBLEMS

5.8   The Fig. 5.21 code listing for `pythagorean()` works, but it only prints the hypotenuse values. Modify the code to print all the sides *a*, *b*, and *c* that are Pythagorean triples, and list the results for sides *a* and *b* between 1 and 20. ®

PRO TIP: NONENGINEERING CAREERS

Many electrical engineering graduates choose to pursue careers other than electrical engineering and find that the quantitative skills they learned as students transfer well. Popular nonengineering careers for EE graduates include the following:

- **Sales:** There are many opportunities for qualified engineers who are more interested in working with people than design. Although these jobs are about building trusting relationships, many require engineering-level knowledge about the system being sold, such as selling substation transformers to provide power to growing communities.

- **Project Management:** This is another people-centric position, and it requires the ability to manage teams of engineers during the development phase of a project. Such positions require far more communication and planning skills than design skills.

- **Military:** Increasing automation, especially in the technology-centric branches of the Air Force and Navy, calls for greater numbers of engineers. There are often special scholarships reserved specifically for engineering majors to help fill this gap.

- **Medicine:** Engineers have higher Medical College Admission Test (MCAT) scores on average than both biology majors and "pre-med" health science majors,[1] and the analytical training they receive gives them advantages in certain specialties, including cardiology, neurology, and radiology.

- **Law:** Math-intensive undergraduate curricula are correlated with Law School Admission Test (LSAT) performance. Engineering majors have among the highest scores on the LSAT,[2] and on average significantly outperform other common majors bound for law school, including language, psychology, social science, liberal arts, and pre-law majors. Patent law recruits heavily from engineering majors: An undergraduate degree in engineering alone is sufficient qualification to take the US Patent Bar Exam.

Opportunities abound both within and outside the engineering profession. Which will you choose?

[1] "MCAT and GPAs for Applications and Matriculants to U.S. Medical Schools by Primary Undergraduate Major, 2019-2020," Association of American Medical Colleges, 2020. www.aamc.org/download/321496/data/factstablea17.pdf.
[2] "Average LSAT Scores by Major." https://magoosh.com/lsat/2016/average-lsat-scores-by-major/.

5.5.2 *Using Nested Loops to Search for* **Best** *Solutions*

Finding exact solutions, as was done in the previous section, is done by testing with the `==` operator. Sometimes exact solutions do not exist, in which case one must settle for finding the best solution. Finding the best solution requires keeping track of the current best solution and checking through every iteration to see if the current computation is better. Error is often computed using code of the form `error = abs(currentComputation − desired)`, since it is the absolute error that needs to be minimized.

For example, if you need to find the value of two standard value resistors from 1 Ω to 1 MΩ in series, whose sum is as close as possible to a given desired value, the program shown in Fig. 5.22 can be used.

```
Python Interpreter                                          >>>
>>> # Best solution search using nested loops
>>>
>>> import numpy as np
>>>
>>> def findClosestSeries(rDesired):
...     """
...     Find two resistors that produce the closest desired sum
...     Input parameters
...         rDesired: the desired series sum
...     returns
...         R1best, R2best: Best resistors producing
...                         closest series sum to rDesired
...     """
...     # Create vR with standard value resistors from 1 to 9.1
...     vR = np.array([1.0, 1.1, 1.2, 1.3, 1.5, 1.6, 1.8, 2.2, \
...                    2.4, 2.7, 3.0, 3.3, 3.6, 3.9, 4.3, 4.7, \
...                    5.1, 5.6, 6.2, 6.8, 7.5, 8.2, 9.1])
...     # Use the backslash, \, to extend to another line
...     # Duplicate vR by multiples of 10 to create all
...     #    standard values of Rs
...     vR = np.concatenate([[0], vR, 10.0 * vR, 100.0 * vR, \
...                          1e3 * vR, 1e4 * vR, 1e5 * vR, \
...                          1e6 * vR])
...     R1best = 0.0 # we haven't yet found the best value of R1
...     R2best = 0.0 # we haven't yet found the best value of R2
...     bestError = 1e12 # our initial error is huge
...     #for R1 in vR:
...     # Loop through to test every possible value of R1
...     for R1 in vR:
...         # Loop through to test every possible value of R2
...         for R2 in vR:
...             Rseries = R1 + R2
...             currentError = np.abs(Rseries - rDesired)
...             if (currentError < bestError):
...                 # we found a better solution
...                 R1best = R1
...                 R2best = R2
...                 bestError = currentError
...     return R1best, R2best
...
>>>
```

Fig. 5.22 Finding the best series combination of two standard resistors to match a desired input value. Two nested `for` loops are used with a test criterion to keep track of the best value.

```
Jupyter Notebook                                              .ipynb
  1  # Best solution search using nested loops
  2
  3  import numpy as np
  4
  5  def findClosestSeries(rDesired):
  6      """
  7      Finds two resistors that produce the closest desired sum
  8      Input parameters
  9          rDesired: The desired series sum
 10      Returns
 11          R1best, R2best: Best resistors producing
 12                          closest series sum to rDesired
 13      """
 14      # Create vR with standard value resistors from 1 to 9.1
 15      vR = np.array([1.0, 1.1, 1.2, 1.3, 1.5, 1.6, 1.8, 2.2, \
 16                     2.4, 2.7, 3.0, 3.3, 3.6, 3.9, 4.3, 4.7, \
 17                     5.1, 5.6, 6.2, 6.8, 7.5, 8.2, 9.1])
 18      # Use the backslash, \, to extend to another line
 19      # Duplicate vR by multiples of 10 to create all
 20      #    standard values Rs
 21      vR = np.concatenate([[0], vR, 10.0 * vR, 100.0 * vR, \
 22                          1e3 * vR, 1e4 * vR, 1e5 * vR, 1e6 * vR])
 23      R1best = 0.0      # we haven't yet found the best value of R1
 24      R2best = 0.0      # we haven't yet found the best value of R2
 25      bestError = 1e12 # our initial error is huge
 26      # Loop through to test every possible value of R1
 27      for R1 in vR:
 28          # Loop through to test every possible value of R2
 29          for R2 in vR:
 30              Rseries = R1 + R2
 31              currentError = np.abs(Rseries - rDesired)
 32              if (currentError < bestError):
 33                  # We found a better solution
 34                  R1best = R1
 35                  R2best = R2
 36                  bestError = currentError
 37      return R1best, R2best
```

Fig. 5.22, cont'd

PRACTICE PROBLEMS

5.9 Change the program listed in Fig. 5.22 to print out the value of two standard-value resistors in parallel closest to a given desired resistor. Test to find the nearest parallel combination of standard resistor values to match $R = 40.5 \ \Omega$. How close is it? ®

5.6 TIMING CODE EXECUTION

To measure the execution times of small segments of Python code, use the `timeit` module. An application of the `timeit` module is illustrated in Fig. 5.23. In this program two 50×50

```
Python Interpreter                                                  >>>
>>> # Timing code execution using the timeit module
>>>
>>> import numpy as np
>>> import timeit
>>>
>>> def timeMatrixMultiplicationSpeed():
...     """
...     Time the speed of matrix multiplication
...         and print out the execution time.
...     Inputs:  None
...     Returns: None
...     """
...
...     A   = np.random.rand(50, 50) # A = random 50x50 matrix
...     B   = np.random.rand(50, 50) # B = random 50x50 matrix
...     tic = timeit.default_timer() # Store initial timer value
...     C   = np.matmul(A, B)        # Matrix multiply A and B
...     toc = timeit.default_timer() # Store final timer value
...     print(toc-tic,'sec Elapsed') # Print elapsed time
...     return None                  # return the object None
...
>>>
```

```
Jupyter Notebook                                                 .ipynb
 1  # Timing code execution using the timeit module
 2
 3  import numpy as np
 4  import timeit
 5
 6  def timeMatrixMultiplicationSpeed():
 7      """
 8      Time the speed of matrix multiplication
 9          and print out the execution time.
10      Inputs:  None
11      Returns: None
12      """
13      A = np.random.rand(50, 50)   # A = random 50x50 matrix
14      B = np.random.rand(50, 50)   # B = random 50x50 matrix
15      tic = timeit.default_timer() # Store initial timer value
16      C   = np.matmul(A, B)        # Matrix multiply A and B
17      toc = timeit.default_timer() # Store final timer value
18      print(toc-tic,'sec Elapsed') # Print elapsed time
19      return None                  # return the object None
```

Fig. 5.23 Timing the speed of matrix multiplication. Using the `default_timer()` function in the `timeit` module, the beginning and end times of program statement(s) can be stored in variables. The difference between the end and beginning times give you the program statement(s) execution time.

random matrix arrays, A and B, are created. The tic variable holds the starting time before the two matrices are multiplied together and the toc variable holds the time just after multiplication. The difference in these two times gives the execution time needed to multiply the two arrays A and B. On an eighth-generation Intel corei7 it took a little over 2 ms to perform the matrix multiplication using this timing program.

5.10 How quickly can Python solve a set of 100 equations with 100 unknowns? This problem was described in Chapter 2. To avoid typing 10,000 test coefficients, create a random problem. First create a 100 × 100 matrix `array` of random numbers and save it in a variable A using the `NumPy random` package. Next, create a 100 × 1 matrix `array` of random numbers and save it in a variable `b`. This part of your code should be of the form

```
# Practice problem 5.10
import numpy as np
import timeit
# Make a 100 x 100 random matrix array
A = np.random.rand(100, 100)
# Make a 100 x 1 random matrix array
b = np.random.rand(100, 1)
```

Using the command

```
V = np.linalg.solve(A, b)
```

time how long it takes to solve the linear equations. ®

5.7 FOR LOOPS USING ELSE, BREAK, AND CONTINUE

5.7.1 The For Else Loop

Upon completion of a `for` loop, an optional `else` code block can be executed. The structure, known as a `for else` loop, is outlined in Fig. 5.24, and the corresponding flowchart is shown in Fig. 5.25. Like the basic `for` loop structure, given in Fig. 5.1, the `for else` loop begins by iterating through each object of a collection. After all the objects of the collection have been assigned to the variable `var`, program flow continues to the indented block of the `else` statement(s). After the `else` code block has finished, program control continues to the next line following the `else` code statement(s).

```
Python Interpreter          >>>
>>> # for else loop structure
>>> for var in collection:
...         statement(s)
... else:
...         statement(s)
...
>>> # Line after for loop
```

```
Jupyter Notebook              .ipynb
1 | # for else loop structure
2 | for var in collection:
3 |         statement(s)
4 | else:
5 |         statement(s)
6 | # Line after for else loop
```

Fig. 5.24 Code structure of a for else loop. After all objects in the collection have been assigned to var, the loop exits and program control moves to the first line of the indented else statement(s).

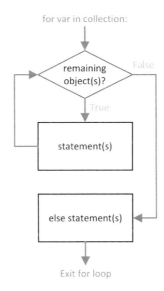

Fig. 5.25 The for else loop flowchart. The for else loop begins by iterating through each object in the collection. When all the objects have been iterated through the for loop program, control moves to the first line of the indented else block. After the else statements have been executed, program control moves to the line following the for else block.

5.7.2 The For Loop Break and Continue Statements

It is sometimes necessary to break out of the middle of a loop before it is completed. For example, this might happen if the loop were searching for some criterion to be met, such as trying to minimize a calculated error below some threshold. Once this is accomplished, the loop has served its purpose. In such a case, you can exit the loop immediately, using the break command. As an alternative, it may be necessary to exit in the middle of an iteration and continue to the next object in the collection. In this case, you use a continue statement. A common use of the break and continue statements is illustrated in the flowcharts shown in Fig. 5.26. An if statement tests some criteria that direct program control to the break or continue statements if it tests True.

(a)　　　　　　　　　　　　　　(b)

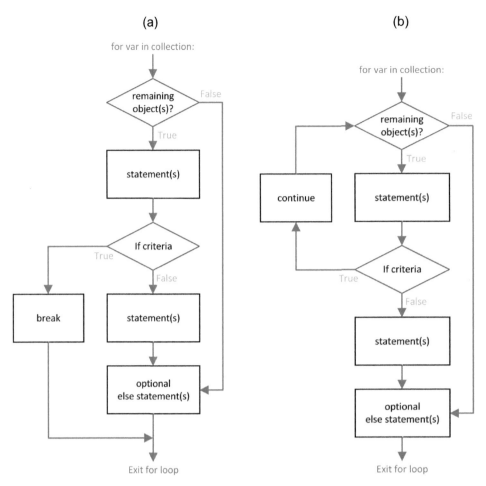

Fig. 5.26 Flowcharts showing `break` and `continue` commands used in a `for` loop. (a) When the `break` command is executed the `for` loop exits and any optional `else` statements are skipped. (b) When a `continue` statement is executed, the remaining part of the `for` loop is bypassed and control returns to the beginning of the `for` loop.

Another example involves searching to see if a given number *n* is prime. One method is to loop over all possible integers between 2 and *n* to see if any are factors of the given number. The moment any factor is found, the loop exits and reports that the number *n* is not prime. If the loop completes with no factors found, the function reports that the number must be prime. But rather than searching for possible factors from 2 through *n*, one only needs to search up to the square root of *n*, since any factor greater than this must be multiplied by a factor less than this. Define a helper function called `isInteger(n)` and then define the main function `isPrime(n)` to locate primes, as shown in Fig. 5.27. This program will run correctly without the `break` statement, but it would be slower. If a factor were found, it would continue to run the loop until the end, rather than returning the result immediately.

```
Python Interpreter                                            >>>
>>> # Test a number to determine if it is prime
>>>
>>> import numpy as np
>>>
>>> # Helper function to test if the input x is an integer
>>> def isInteger(x):
...     # is_integer will be True if x is an integer
...     # otherwise is_integer will be False
...     is_integer = (x==np.floor(x))
...     return is_integer
...
>>> # Main function to test if the input n is a prime
>>> def isPrime(n):
...     # Assume is_prime is prime then check
...     is_prime = True
...     # pos_factor cycles through all possible factors
...     for pos_factor in np.arange(2,np.floor(np.sqrt(n))+1):
...         if isInteger(n/pos_factor): # Is this a factor of n?
...             is_prime = False          # Then n is not prime
...             break                     # break out of the loop
...     return is_prime
...
>>>
```

```
Jupyter Notebook                                           .ipynb
 1  # Test a number to determine if it is prime
 2
 3  import numpy as np
 4
 5  # Helper function to test if the input x is an integer
 6  def isInteger(x):
 7      # is_integer will be True if x is an integer
 8      # otherwise is_integer will be False
 9      is_integer = (x==np.floor(x))
10      return is_integer
11
12  # Main function to test if the input n is a prime
13  def isPrime(n):
14      # Assume is_prime is prime then check
15      is_prime = True
16      # pos_factor cycles through all possible factors
17      for pos_factor in np.arange(2,np.floor(np.sqrt(n))+1):
18          if isInteger(n/pos_factor): # Is this a factor of n?
19              is_prime = False          # Then n is not prime
20              break                     # break out of the loop
21      return is_prime
```

Fig. 5.27 Test a number to determine if it is prime. A helper function `isInteger()` is used with the main function `isPrime()` to test if the input is a prime number.

PRACTICE PROBLEMS

5.11 Change the code for the function `isPrime()` listed in Fig. 5.27 to check to make sure that the program has not been running for over 1 second. Do this by starting a timing operation at the beginning of the code using `timeit.default_timer()`, and check the elapsed time inside the loop again with timeit.default_timer(). Use the break statement to exit if the code has been executing for over 1 second, and if so exit and return −1. This is a useful technique for exiting automatically when a simulation does not converge on an answer. Search the internet for very large prime numbers and try them. At the time of writing, the prime number 999,999,000,001 took more than a second to process. ©

5.8 WHILE LOOP

The for loop iterates a definite number of times known in advance, unless a break statement ends it early. The while loop, on the other hand, tests a logical expression and will iterate indefinitely as long as the logical expression tests True. The while loop terminates when the logical expression tests False. The basic while loop code structure and flowchart are shown in Figs. 5.28 and 5.29, respectively.

5.8.1 Using a Counter in a While Loop

The iteration of a while loop can be controlled by introducing a counter variable. For example, the program listing in Fig. 5.30 initializes the counter variable i to zero and then increments it during each while loop iteration. When the counter i is no longer less than four, it terminates. A variation on this program would be to start the counter at some value and decrement it during each loop iteration. The loop can be terminated when the variable counts down to a value that sets a logical expression False.

Fig. 5.28 Basic while loop code structure. The while loop is entered each time the logical expression evaluates to True. When the logical expression is False when it is evaluated, the while loop terminates.

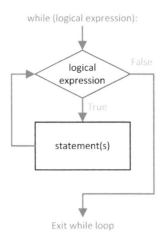

Fig. 5.29 Basic while loop flowchart. The while loop is entered each time the logical expression tests True.

```
Python Interpreter          >>>
>>> # while loop using counter
>>>
>>> i = 0 # Set counter to 0
>>> while i < 4:
...        print('i =', i)
...        i = i + 1
...

i = 0
i = 1
i = 2
i = 3
```

```
Jupyter Notebook          .ipynb
1  # while loop using counter
2
3  i = 0 # Set counter to 0
4  while i < 4:
5        print('i =', i)
6        i = i + 1
7

i = 0
i = 1
i = 2
i = 3
```

Fig. 5.30 Using a counter variable in a while loop. Before the while loop starts, the variable i is set to zero. During each iteration, the variable i is incremented. When the counter variable i reaches 4 and the logical expression tests False, the loop exits.

5.8.2 The While Else Loop

Like the for loop, the while loop can include an optional else statement. After the while loop terminates, it will execute the statements included in the else block. The code framework for a while else loop and its associated flowchart are shown in Figs. 5.31 and 5.32, respectively.

A simple example of a while else loop is shown in Fig. 5.33. Notice that upon completion of the while loop, the else block print statement is executed. Although it is possible to include else statements in a while loop, it is not as common as the for else loop.

```
Python Interpreter          >>>
>>> # while else loop
>>> # structure
>>>
>>> while logical_expression:
...     statement(s)
... else:
...     statement(s)
...
>>>
```

```
Jupyter Notebook            .ipynb
1  # while else loop
2  # structure
3
4  while logical_expression:
5      statement(s)
6  else:
7      statement(s)
8
9
```

Fig. 5.31 Basic `while` loop code structure with an `else` statement block. The `while` loop is entered each time the logical expression evaluates to `True`. When the expression becomes `False`, the `while` loop executes the `else` block statement(s) and then terminates.

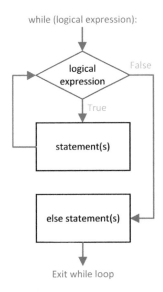

Fig. 5.32 Flowchart for `while else` loop. Once the `while` loop has completed, an optional `else` statement block can be included.

```
Python Interpreter          >>>
>>> # while else loop example
>>>
>>> i = 0
>>> while i < 4:
...     print('i =', i)
...     i = i + 1
... else:
...     print('Done.')
...
i = 0
i = 1
i = 2
i = 3
Done.
```

```
Jupyter Notebook            .ipynb
1  # while else loop example
2
3  i = 0
4  while i < 4:
5      print('i =', i)
6      i = i + 1
7  else:
8      print('Done.')
9

i = 0
i = 1
i = 2
i = 3
Done.
```

Fig. 5.33 Code example for `while else` loop. When the counter reaches 4, the `while` loop exits and the single `print` statement in the `else` block is executed.

5.8.3 *The While Loop Break and Continue Statements*

The break and continue statements can also be included in a while loop. When a break statement occurs, the while loop ends, bypassing any optional else statements, and control passes to the code following the while loop. When a continue statement is executed in a while loop, program control returns to the while loop logical expression. Fig. 5.34 shows the flowcharts associated with the break and continue statements when they are used in a while loop.

Study the example shown in Fig. 5.35 and make sure you understand the how the continue and break statements lead to the output shown.

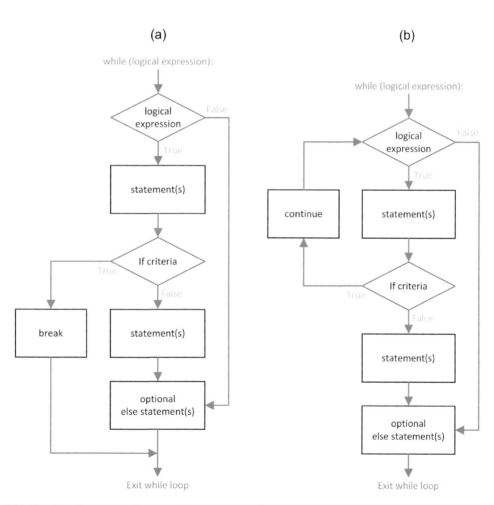

Fig. 5.34 Flowcharts for while loops containing break and continue statements. (a) A break statement exits the loop and bypasses the optional else statements. (b) When a continue command is executed, program control returns to the beginning of the loop.

```
Python Interpreter          >>>        Jupyter Notebook            .ipynb
>>> # while loop                        1 | # while loop
>>> #   with break & continue           2 | #   with break & continue
>>>                                     3 |
>>> i = 0                               4 | i = 0
>>> while i < 4000000:                  5 | while i < 4000000:
...     i = i + 1                       6 |     i = i + 1
...     if i == 2:                      7 |     if i == 2:
...         continue                    8 |         continue
...     print('i =', i)                 9 |     print('i =', i)
...     if i == 4:                     10 |     if i == 4:
...         break                      11 |         break
... else:                             12 | else:
...     print('else block.')          13 |     print('else block.')
...                                    14 |
i = 1                                   i = 1
i = 3                                   i = 3
i = 4                                   i = 4
```

Fig. 5.35 Including `break` and `continue` commands in a `while` loop. When a `continue` command is executed, the remainder of the loop is bypassed and control returns to the beginning of the `while` loop. A `break` statement exits the loop altogether, bypassing any optional `else` statements.

5.8.4 Terminating an Infinite Loop

At some point you will inevitably write a `while` loop that does not terminate. This is known as an infinite loop. For example, if you forget to increment your counter using the code segment

```
>>> i=0
>>> while i<4:
...        print('i=',i)
...
```

the `while` loop will not end and 0s will be repeatedly printed. Using `Ctrl+C` or `Ctrl+D` will terminate an infinite loop on some systems but not all. In some command windows it may be necessary to close the window and start again. In a Jupyter Notebook you can terminate an infinite loop by interrupting the kernel. Choose Interrupt from the Kernel drop-down menu, as illustrated in Fig. 5.36. You should take the time to understand how to terminate an infinite loop before you run one.

5.8.5 Infinite Series Approximations Using a While Loop

As an application of the `while` loop, consider the calculation of the infinite series $\frac{1}{2} + \frac{1}{4} + \frac{1}{8} + \dots$. This series can be approximated to a given precision x by adding successive terms until the term value becomes less than x. A Python function to evaluate this series to a precision x is of the form

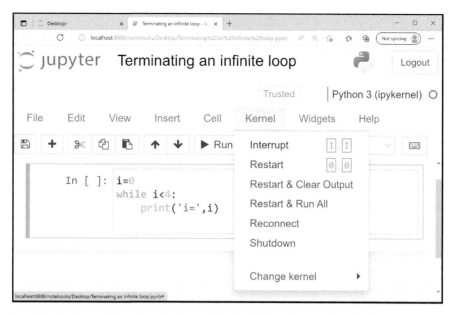

Fig. 5.36 Terminating an infinite loop in a Jupyter Notebook. Select Interrupt from the Kernel pulldown menu.

```
>>> def geometricSeries(x):
...     """ Loop until new additions are < x """
...     term = 1.0
...     seriesSum = 0.0
...     while (term >= x):
...         term = term / 2.0
...         seriesSum = seriesSum + term
...     return seriesSum
...
```

PRACTICE PROBLEMS

5.12 Write a function to find the value of the infinite series $\frac{1}{3} + \frac{1}{9} + \frac{1}{27} + \cdots$ to a precision x using a while loop structure. Report the result for summing each term until the last term added is smaller than 0.00001. ®

5.9 GROWING VERSUS PREALLOCATING VECTOR ARRAYS

Often a function will return an `array` with objects calculated inside a loop. In this case, the `array` may be either initialized to the full size with the `NumPy zeros()` function and then filled inside the loop or else grown to size inside the loop using the `NumPy append()` function with its length increasing with every loop iteration. The former method is much more efficient and is accomplished by reserving the required memory for the `array` before the loop begins. For example,

```
>>> import numpy as np
>>> result = np.zeros(1000)
```

The latter, less efficient method is to grow the size of the `array` in the middle of the loop using the `NumPy append()` function. For small `arrays` outside of a loop, however, the `NumPy append()` function can add a new value to an `array`. For example,

```
>>> import numpy as np
>>> v = np.array([1, 2])
>>> v = np.append(v, [3])
>>> print(v)
[1 2 3]
```

If the resulting vector size is known before the loop begins executing, as is often the case, use the former method. Only if the resulting vector length is not known in advance should the latter method be used. Figs. 5.37 and 5.38 provide examples of preallocation and growing `array` memory in a loop. Study them carefully so you understand when to use preallocation.

DIGGING DEEPER
The reason that vector arrays should be initialized with values, such as zeros, rather than grown as needed, is the way Python manages memory. Consider a vector array of length 100 that needs to be grown to length 101 to accommodate a new value. Python may have to first create internally an entirely new length 101 vector array. The original 100 values must then be copied over to it, and the additional single value must be saved. Compare this with the faster alternative of preallocating memory for the final array once and then replacing the initialized values with the calculated values.

```
Python Interpreter                                                    >>>
>>> # Good programming example using a preallocated array
>>>
>>> import numpy as np
>>>
>>> def goodProgram(n):
...     """Preallocate v of length n and return v"""
...     v = np.zeros(n)  # Preallocate v with zeros of length n
...     for i in range(n):
...         # Load v with data
...         v[i] = np.random.rand() + np.sin(2.0*i)
...     return v  # Return v to calling program
...
>>>
```

```
Jupyter Notebook                                                   .ipynb
 1  # Good programming example using a preallocated array
 2
 3  import numpy as np
 4
 5  def goodProgram(n):
 6      """Preallocate v of length n and return v"""
 7      v = np.zeros(n)  # Preallocate v with zeros of length n
 8      for i in range(n):
 9          # Load v with data
10          v[i] = np.random.rand() + np.sin(2.0*i)
11      return v  # Return v to calling program
```

Fig. 5.37 Good programming example using array preallocation. Preallocating your arrays before filling them in a loop in most cases produces faster code and less memory management.

```
Python Interpreter                                                    >>>
>>> # Bad programming example without preallocating array
>>>
>>> import numpy as np
>>>
>>> def badProgram(n):
...     """ Grow v to length n and return v """
...     v = np.array([]) # Start with an empty array
...     for i in range(n):
...         # Load v with data
...         newval = np.random.rand() + np.sin(2.0*i)
...         v = np.append(v, [newval]) # Append newval to v
...     return v # Return v to calling program
...
>>>
```

```
Jupyter Notebook                                                   .ipynb
 1  # Bad programming example without preallocating array
 2
 3  import numpy as np
 4
 5  def badProgram(n):
 6      """ Grow v to length n and return v """
 7      v = np.array([]) # Start with an empty array
 8      for i in range(n):
 9          # Load v with data
11          newval = np.random.rand() + np.sin(2.0*i)
12          v = np.append(v, [newval]) # Append newval to v
13      return v # Return v to calling program
```

Fig. 5.38 Growing arrays in a for loop. When you know the dimensions of your array before entering the loop, appending values during each iteration is a slower process than preallocation. Whenever possible, you should preallocate your arrays.

There are two types of mathematics electrical engineers do with `False` and `True` or 0 and 1; **logical** and **binary**.

Logical or Boolean Math

Logical or Boolean math use numbers to represent `True` (1) and `False` (0). The logical operations of `and`, `or`, and `not` model the effect of their counterpart logic gates and, or, and not. The symbols for the and, or, and not gates are shown in Fig. 5.39.

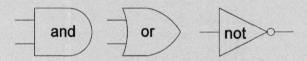

Fig. 5.39 And, or, and not gate symbols.

As an example, find the output of the digital circuit shown in Fig. 5.40 using Python.

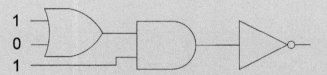

Fig. 5.40 Digital circuit consisting of an or-and-not gate sequence.

The output can be found by substituting the Python logical operators as shown in Fig. 5.41.

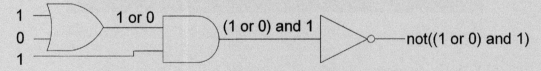

Fig. 5.41 Finding the digital output of an or-and-not gate sequence.

The Python code to calculate the answer given the input, therefore, is `not((True or False) and True)`, which evaluates to `False`.

Binary Math: `bin()` **and** `int()`

Binary mathematics involves calculations with base 2 numbers, but Python shows all numbers in base 10. To convert to base 2, use `bin()`.
For example, to convert 77 to base 2, type

```
>>> bin(77)
```

which returns `0b1001101`.

The prefix indicates that the `0b` number is a binary representation. To remove the `0b` prefix, use

```
>>> bin(77).replace("0b", "") returns 1001101
```

To convert back into base 10, use `(int)(b,2)` and put string markers around the binary value:

```
>>> (int)("0b1001101",2)
```

Which returns `77`.

Note that not including the 0b prefix also gives the correct decimal value:

```
>>> (int)("1001101",2)
```

also returns `77`.

As an example, perform the following operation in Python:
$10011_2 \times 1001_2 + 1010_2$.

```
>>> (int)("10011",2) * (int)("1001",2) + (int)("1010",2)
```

This returns 181 (in base 10). To see the result in binary, use

```
>>> bin(181)
```

This returns

`'0b10110101'` (in base 2).

5.13   Use Python to calculate the logic block output shown in Fig. 5.42. ©®

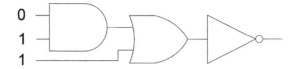

Fig. 5.42 And-or-not gate sequence.

5.14   Use Python to evaluate the binary math problem ($1001_2 +$ 1110_2) $\times$ 10011_2 and report the answer in binary. ®

PRO TIP: FUNDAMENTALS OF ENGINEERING

An earlier Pro Tip discussed the advantages of taking the optional step of earning the professional engineering license. The first step in becoming a PE is to pass the Fundamentals of Engineering (FE) exam. This test is designed to be taken during the senior year or by recent graduates of an engineering program. There are versions of the FE exam for electrical, mechanical, civil, and other engineering majors; the FE exam includes topics common to all these majors, such as mathematics, probability, ethics, engineering economics, and topics unique to the selected engineering discipline. The FE exam for EEs also includes circuit analysis, signal processing, electromagnetics, digital systems, and programming.

The exam is composed of two three-hour computer-graded segments, which are taken with a break for lunch. Certain models of scientific nonprogrammable calculators are authorized, and a book containing reference equations is provided. You can find out more about the exam, how to purchase a copy of the reference equation book, and how to register for it by visiting NCEES.org, the administering testing agency. Usually, a professor or advisor in your department will also have information on the exam. Your department or IEEE student branch might even run review sessions, so reach out to identify your local resources.

Although scores are reported, it is essentially a pass/fail exam since its only purpose is to serve as one of several steps required to become eligible to take the PE exam. In past years, roughly 70% of electrical and computer engineering test-takers earned passing scores, and one can retake the exam as many times as required, up to once in every two-month testing window, and up to three times per year.

Very few undergraduate majors have a national licensing examination as engineers do. Although most electrical engineering careers do not require the PE, it is difficult to determine as a student where your particular career path will lead. Why not take advantage of the fact the national licensure exists for electrical engineers and gain maximum career flexibility and a distinguishing resume line by planning to take the FE as a senior?

COMMAND REVIEW

for loop

```
for var in collection:
    statement(s)
for var in collection:
    statement(s)
else:
    else statement(s)
break
continue
```

Basic `for` loop structure. Iterate over `collection` objects.
Statement(s) executed during each `for` loop iteration.
The `for` loop structure with `else` statement(s).
Statement(s) executed during each `for` loop iteration.

Statement(s) executed when `for` loop ends with no `break`.
Terminate `for` loop and bypass optional `else` statement(s).
Skip the rest of the `for` loop and return to the beginning of the loop.

while loop

```
while (logic statement):
    statement(s)
while (logical statement):
    statement(s)
else:
    else statement(s)
break
continue
```

Iterate `while` loop when logical statement is `True`.
Statement(s) executed during each `while` loop iteration.
A `while` loop can have optional `else` statement(s).

`else` statement(s) execute when loop ends with no `break`.
Terminate `while` loop and pass optional `else` statement(s).
Skip the remainder of the loop and return to loop beginning.

Terminating an infinite loop

Ctrl-C, Ctrl-D, or Interrupt Kernal

Be able to terminate an accidental infinite loop on your system!

Preallocation

```
import numpy as np
result = np.zeros(N)
```

To use `zeros()` make sure you import the `NumPy` package.
Preallocate arrays with `N zeros` before using them in loops.

Binary functions

```
bin(72)
bin(72).replace('0b', '').
(int)('0b1001000',2)
(int)('1001000',2)
(int)('10011',2) * (int)('1001',2) + (int)('1010',2)
```

Convert a base 10 value into base 2, `'0b1001000'`.
Convert a base 10 value into base 2 with no prefix, `'1001000'`
Convert a base 2 number, in quotation marks, into base 10, 72.
Convert a base 2 number, in quotation marks, into base 10, 72.

Binary operation $10011_2 \times 1001_2 + 1010_2 = 181$.

LAB PROBLEMS

©=Write only the Python command(s).

®=Write only the Python result.

Solutions to all starred (*) problems are available on the Elsevier website (refer to page xix for the link).

5.1* Write a Python program that takes a `NumPy array t` and returns a `NumPy array v`, according to the formula

$$v(t) = \begin{cases} 1 + t, & t \le 0 \\ 2e^{-t} - 1, & 0 < t \end{cases} \tag{5.5}$$

where *t* is the time in seconds and *v(t)* is the voltage in volts. Use a `for` loop in your solution. The structure of the function should be of the form

```
def labProblem5_1(t):
    # statements
    return v
```

a) Test it by verifying that

```
import numpy as np
t=np.array([-5, 1])
labProblem5_1(t)
```

returns the `NumPy array [-4., -0.26424112]`. ©®

b) Add code to plot and label `v` vs `t` for $-2 \le t \le 2$. ©®

5.2 A compander is a device that passes audio signals from -1 V to 1 V without change, but reduces the gain of higher-magnitude signals, in part to prevent large transient short-length signals that can saturate the system. You will learn more about companders in communication courses. An example of a compander is the rule

$$v_{out}(t) = \begin{cases} \dfrac{1}{2}v_{in} - \dfrac{1}{2}, & -2 \le v_{in} < -1 \\ v_{in}, & -1 \le v_{in} < 1 \\ \dfrac{1}{2}v_{in} + \dfrac{1}{2}, & 1 \le v_{in} \le 2 \end{cases} \tag{5.6}$$

Note that the horizontal axis, or "x" axis, is v_{in} and the vertical axis, or "y" axis, is v_{out}. Write a Python program that takes a vector array of values for v_{in} and returns the associated vector array v_{out}. It should have the function definition

```
def labProblem5_2(vin):
    # statements
    return vout
```

a) Test the code by verifying that

```
import numpy as np
vin = np.array([-2, 0, 2])
labPoblem5_2(vin)
```

returns the `NumPy array [-1.5, 0., 1.5]`.

b) Add code to plot and label `vout` vs `vin` for $-2 \leq vin \leq 2$. ©®

5.3 Write a Python function that creates a triangular waveform that oscillates between 0 V and 1 V, starting at $t = 0$. The function should take three arguments: the number of points computed, the triangular wave period, and the length of time that the triangle waveform is to be computed. It should return two arguments, the time vector `t` and the voltage vector `v`, so that `axes` `plot(t,v)` plots the result. The function definition should be of the form

```
import numpy as np
def labProblem5_3(nPoints, period, stopTime):
    t = np.linspace(0.0, stopTime, nPoints)
    # statements
    return t, v
```

There are many ways to write this function. One way is to calculate `v[i]` inside a loop indexed by `i` for every time value in `t`. For each value of `t` being inspected, first find how long it has been since the beginning of the last period, and then offset `t` by this amount. Check that the function works correctly. The input

```
import matplotlib.pyplot as plt
# use %matplotlib inline if using a Jupyter Notebook
t, v = labProblem5_3(21, 2, 10)
```

```
fig, ax = plt.subplots()
ax.plot(t,v)
ax.set_xlabel("Time, t (s)") # Label, symbol (units)
ax.set_ylabel("Voltage, v (V)"); # Use a ; to suppress output
# Use plt.show() if using Python interpreter
```

should produce the plot shown in Fig. 5.43.

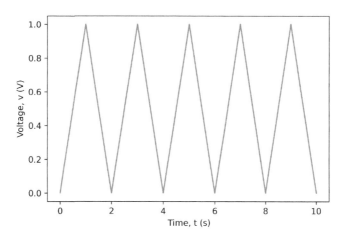

Fig. 5.43 Triangular waveform oscillating between 0 and 1 (Lab Problem 5.3).

5.4 In digital signal processing, the Tukey window is defined by the function

$$
y(x) = \begin{cases} \dfrac{1}{2} + \dfrac{1}{2}\cos\left(\dfrac{2x\pi}{\alpha N} - \pi\right), & 0 \leq x < \dfrac{\alpha N}{2} \\[2mm] 1, & \dfrac{\alpha N}{2} \leq x < N - \dfrac{\alpha N}{2} \\[2mm] \dfrac{1}{2} + \dfrac{1}{2}\cos\left(\dfrac{2x\pi}{\alpha N} - \dfrac{2\pi}{\alpha} + \pi\right), & N - \dfrac{\alpha N}{2} \leq x \leq N \end{cases} \quad (5.7)
$$

This window is also only defined for integers x in the range $0 \leq x \leq N$. Write a function that takes arguments N and α and checks to make sure N is a positive integer and that α is between 0 and 1. The function should return two arguments, x and the Tukey window function $y(x)$, so that the `axes plot(x, y)` function can display the result. If either N or α is not valid, it returns the error value -1 for both x and y.

The function definition and a few lines of the code should be of the form

```
import numpy as np
def labProblem5_4(N, alpha):
    x = np.arange(0,N+1)
    for i in np.arange(0,N+2):
        # statement(s)
    return x, y
```

Check that the function works correctly. The code segment

```
x, y = labProblem5_4(100, 0.5)
fig, ax = plt.subplots()
ax.plot(x, y)
ax.set_title("Tukey Window, alpha = 0.5")
# Use plt.show() if using the Python interpreter
```

should produce the plot shown in Fig. 5.44.

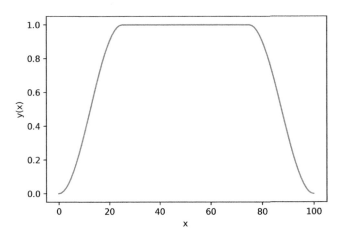

Fig. 5.44 Tukey window using an α value of 0.5 (Lab Problem 5.4).

5.5* In the "Nested Loops" section 5.5, Fig. 5.20 defines a function called findResistors() that outputs all combinations of four resistors taken two at a time. However, it does not recognize that R1 = 10, R2 = 22 is the same combination as R1 = 22, R2 = 10. Modify the code to avoid listing duplicates such as these. This requires some thought. It may help to find the pattern by crossing out the repeats in the output table. The function definition line should be

```
def labProblem5_5():
```

5.6 Section 5.5.2, "Using Nested Loops to Search for Best Solutions," has code to find the two standard-value resistors whose value in series is the nearest to a given desired value. The associated Practice Problem 5.9 asks you to modify the code to return the two standard-value resistors whose value in parallel is the nearest to a given desired value. Modify the code so that it returns the single resistor, or set of two resistors in series or parallel, that is closest to the given value. The function definition should be of the form

```
def labProblem5_6(rDesired):
    # statement(s)
    return R1best, R2best, result, Rtype
```

The result is the equivalent resistance computed with `R1best` and `R2best`, which shows both how close the match to `rDesired` is and whether `R1best` and `R2best` should be connected in series or parallel. If the closest match is a single resistor, then `R2best` should be 0. Test the code by verifying that one can create an 8.5 Ω resistor by placing a 1 Ω and a 7.5 Ω resistor in series; equivalently,

```
r1, r2, res, rtype = labProblem5_6(8.5)
```

evaluates to

```
r1 = 1.0, r2 = 7.5, res = 8.5, rtype = "Series"
```

Similarly, verify that the best way to create a 2.555 Ω resistor is by placing a 4.7 and 5.6 Ω resistor in parallel; that is,

```
r1, r2, res, rtype = labProblem5_6(2.555)
```

evaluates to

```
r1 = 4.7, r2 = 5.6, res = 2.5553, rtype = "Parallel"
```

5.7* A Pythagorean triple, as previously discussed, is a rectangle in two dimensions with integer sides and an integer diagonal. It must satisfy the equation $x^2 + y^2 = d^2$ — for example, 3, 4, and a diagonal of 5, since $3^2 + 4^2 = 5^2$. In three dimensions, this becomes a rectangular prism, or box, with three integer sides and an integer diagonal connecting opposite corners. It must satisfy the equation

$x^2 + y^2 + z^2 = d^2$. For example, 1, 2, 2, and a diagonal of 3, since $1^2 + 2^2 + 2^2 = 3^2$. Write a program to find one such set of four numbers, in which each of the sides and diagonal are at least 5 and not more than 11. Hint: Three sets of nested loops are needed, one for each variable x, y, and z. It takes no arguments and returns four different scalars x, y, z, and d with any single solution. There is more than one solution. The function definition and a few lines of code should be of the form

```
import numpy as np
def labProblem5_7():
    for x in np.arange(5,12):
        # statement(s)
    return xPyth, yPyth, zPyth, dPyth
```

5.8 A single roll of a six-sided die yields an integer from 1 to 6, all equally likely. Two six-sided dice summed will yield a value between 2 and 12, and all values will not be equally likely since there are many ways to roll a total of 6 (1 + 5, 2 + 4, 3 + 3, 4 + 2, 5 + 1) but only one way to roll a 2 (1 + 1). Write a function called `labProblem5_8roll()` using a for loop that takes a number of rolled six-sided dice and sums up their values. For example, `labProblem5_8roll(4)` provides the sum of a virtual roll of four six-sided dice. Its output will be different between calls.

a) First, create the function `labProblem5_8roll()` to do this.

b) Second, write a second function called `labProblem5_8()` that calls `labProblem5_8roll()`. The function `labProblem5_8()` takes two numbers: a number of six-sided dice to be rolled, called `nDice`, and a number of trials, called `nTrials`. It returns a vector of length nTrials, each value of which corresponds to an output of `labProblem5_8roll()`. For instance, if `nDice` is 2 and `nTrials` is 8, one run might randomly return the vector [4 2 5 6 8 4 12 7], representing eight different sums of 2 six-died dice.

c) Provide a histogram of the sum of 10 six-sided dice computed for 100,000 different trials. The histogram should look roughly like a Gaussian curve, even though it is composed of sums of uniformly distributed random numbers. This tendency of sums of random variables to form a Gaussian curve is called the central limit theorem. Provide the histogram image in a separate Word file.

The function declarations and some code lines are

```
import numpy as np

def labProblem5_8(nDice, nTrials):
    results = np.zeros(nTrials) # Preallocate
    for i in np.arange(nTrials):
        # statement(s)

def labProblem5_8roll(nDice):
    total = 0
    # statement(s)
    return total
```

where the `# statement(s)` indicate one or more lines of unwritten code.

5.9 Filters are common circuits that attenuate a signal at certain frequencies but pass it at others. You were introduced to low-pass filters in the Tech Tip *Low-Pass Filters* in Chapter 3. Another type of low-pass filter is shown in Fig. 5.45. It passes low frequencies but substantially reduces frequency components higher than f_c, where f_c is given by

$$f_c = \frac{1}{2\pi\sqrt{R_1 R_2 C_1 C_2}}$$

(5.8)

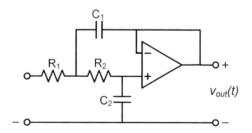

Fig. 5.45 Low-pass filter (Lab problem 5.9).

a) Write a function that computes f_c given values of R_1, R_2, C_1, and C_2. Call this `labProblem5_9a()`. To test it, verify that when the function is called with two identical 10 kΩ resistors and two identical 1 μF capacitors, the cutoff frequency f_c returned is 15.916 Hz.

b) Let the resistors be 10 kΩ with 5% tolerance (meaning each could be anywhere from 9.5 to 10.5 kΩ) and the capacitors be 1 μF with a 10% tolerance. Write a second function called `labProblem5_9b()` that calls `labProblem5_9a()`. The function `labProblem5_9b()` takes a number N representing how many times the circuit will be simulated using different component values. For each of the N trials, it generates random values for R_1, R_2, C_1, and C_2 that are within the components' tolerances, calls `labProblem5_9a()` to determine the corresponding cutoff frequency fc, and stores this value in an N-length vector vFc, which it returns.

c) Provide a histogram of the resulting frequencies fc stored in vFc for 100,000 trials. The histogram gives a sense of the range of values of cutoff frequency fc that will be encountered given real-world components. The function declarations and some code lines are

```
def labProblem5_9a(R1, R2, C1, C2):
    # statements
    return fc
def labProblem5_9b(N):
    vFc = np.zeros(N) # Preallocate
    for i in range(N):
        # statements
    return vFc
```

where the `# statement(s)` indicate one or more lines of unwritten code.

5.10* A voltage divider, previously discussed in the Chapter 2 Tech Tip *Voltage Dividers*, is given by the circuit schematic diagram shown in Fig. 5.46. The circuit input voltage, V_{in}, is divided between the two resistors, so that the voltage across R1 is

$$V_1 = V_{in}\frac{R_1}{R_1 + R_2} \tag{5.9}$$

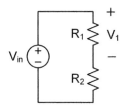

Fig. 5.46 Voltage divider (Lab problem 5.10).

a) Write a function that computes $v1$ given values of Vin, $R1$, and $R2$. Call this function `labProblem5_10Divider()`. To test it, verify that when called with $V_{in} = 10$ V, $R_1 = 1$ kΩ, and $R_2 = 10$ kΩ, it returns a value, rounded to four decimal places, equal to 0.9091 V.

b) Let $V_{in} = 10$ V exactly, $R_1 = 1$ kΩ $\pm$ 5%, and $R_2 = 10$ kΩ $\pm$ 5%. The 5% resistor tolerance means the 1 kΩ could be anywhere from 1 kΩ $-$ 5%=0.95 kΩ to 1 kΩ +5% = 1.05 kΩ. Write a second function called `labProblem5_10()` that calls `labProblem5_10Divider()`. The function takes as input a number N representing how many times the circuit will be simulated using different random component values. For each of the N trials, it generates random values for R_1 and R_2 that are within the component's tolerances, calls `labProblem5_10Divider()` to determine the corresponding voltage V_1, and stores this value in an N-length vector $v1$, which it returns.

c) Provide a histogram of the resulting voltage stored in $v1$ for 100,000 trials. The histogram gives a sense of the range of values of voltage V1 that will be encountered given real-world components.

The function declarations and some code lines are

```
def labProblem5_10(N):
    v1 = zeros(N) # preallocate
    for i in range(N):
        # statements
    return v1


def labProblem5_10Divider(Vin, R1, R2):
    # statement
    return v1
```

where the commented statements indicate one or more lines of unwritten code.

5.11 A fault-tolerant computer has three redundant cores. Each core has a base 20% chance of failing. If any core fails, then recheck both the remaining cores a second time to see if they fail — now they will fail with a 30% likelihood. The fault-tolerant computer will fail if more than 1 core fails.

a) Write a function that computes whether, for a particular trial, the fault-tolerant computer fails. Do this by generating a random value for each core to determine if it fails. If any one core fails, then generate new random values for each of the remaining cores to see if they fail. If they do, the entire system fails and the function should return a 0; otherwise, the system works and the function returns a 1. Call this function `labProblem5_11Trial()`.

b) Write a second function called `labProblem5_11()` that calls `labProblem5_11Trial()`. The function takes a number N representing how many times the fault-tolerant computer system will be simulated. For each of the N trials, it calls to determine if the computer simulation worked and stores this value in an N-length vector v, which it returns.

c) Report the mean and standard deviation of the number of times the computer system worked over 100,000 simulations. Hint: Use the `NumPy` `mean()` and `NumPy` `std()` functions. The mean should be greater than the standard deviation, and both should be less than 1.

The function declarations and some lines are

```
def labProblem5_11(N):
    v1 = zeros(N) # preallocate vector array
    for i in range(N):
        # statements
    return v1

def problem11Trial():
    # statement
    return works
```

where the commented statements indicate one or more lines of unwritten code.

5.12 One method of generating a square wave that alternates between 5 V and 0 V is to use a 555-timer integrated circuit, as shown in Fig. 5.47. This method will be discussed in greater detail in Chapter 6, but this problem only requires the equation governing the time t_{low} that each square wave is 0 V, which is $t_{low} = 0.693\ C\ (R_1 + R_2)$.

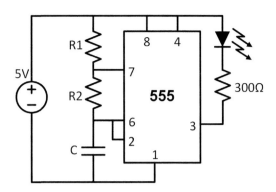

Fig. 5.47 Timer circuit using a 555 chip (Lab Problem 5.12).

a) Write a function that computes t_{low} given values of R_1, R_2, and C. Call this function `labProblem5_12tlow()`. Test it by verifying that when called with $R_1 = 1$ kΩ, $R_2 = 1$ kΩ, and C = 1 μF, it returns a value of 1.386 ms (i.e., 0.001386).

b) Let $R_1 = 1$ kΩ ± 5%, $R_2 = 1$ kΩ ± 5%, and C = 1 μF ± 10%. The 5% resistor tolerance means the 1 kΩ resistor could have a resistance anywhere from 1 kΩ − 5% = 0.95 kΩ to 1 kΩ + 5% = 1.05 kΩ. Capacitors usually have looser tolerances than resistors. Write a second function called `labProblem5_12()` that calls `labProblem5_12tlow()`. The function `labProblem5_12()` takes a number N representing how many times the circuit will be simulated using different component values. For each of the N trials, it generates random values of R_1, R_2, and C that are within the components' tolerances, calls `labProblem5_12tlow()` to determine the corresponding time t_{low}, and stores this value in an N-length vector v, which it returns.

c) Provide a histogram of the resulting t_{low} times stored in v for 100,000 trials. The histogram gives a sense of the range of times t_{low} that will be encountered given real-world components.

The function declarations and some lines of code are

```
import numpy as np
def labProblem5_12(N):
    v1 = np.zeros(N) # preallocate vector array
    for i in range(N):
        # statements
```

where the commented statements indicate one or more lines of unwritten code.

5.13* Most functions encountered in electrical engineering are well-behaved, which makes exceptions all the more interesting. A Weierstrass function is one such exception. This is a continuous function since it can be drawn without lifting the pen from the paper. However, it is filled with so many abrupt changes in direction that it is nowhere differentiable; that is, it has no definite slope anywhere. One curve in this family of functions is defined in the limit as $N \to \infty$ of the equation

$$y(t) = \sum_{n=0}^{N} \left(\frac{1}{2}\right)^n \cos(7^n \pi t) \tag{5.10}$$

a) Write a Python function that takes argument N and returns 10,000 pairs of points $\{t, y\}$ that plot the function over the range $-1 \le t \le 1$. The first few lines of the function definition are

```python
import numpy as np
def labProblem5_13(N):
    t = np.linspace(-1, 1, 1e4)
    y = np.zeros(1e4) # preallocate vector array
    for i in range(1e4):
        tCurrent = t(i)
            for n in range(N+1):
            # statement(s)
    return t, y
```

b) Plot the result for $N = 100$. The plot should look jagged and self-similar when zoomed in a little. Python limited precision prevents making N arbitrarily large, since the 7^n term becomes large quickly, but in the limit the graph would be jagged and self-similar, regardless of the zoom level.

5.14 A common waveform encountered in power supply design is the full-wave nonprecision rectified sinusoid, shown in Fig. 5.48.

One cycle of this waveform is described by the equation

$$y(t) = \begin{cases} \sin\left(\dfrac{10}{9}\pi t\right), & 0 \le t \le 0.9 \\ \\ 0, & 0.9 < t \le 1 \end{cases} \tag{5.11}$$

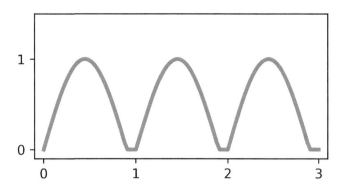

Fig. 5.48 Full-wave rectifier (Lab Problem 5.14).

a) Write a function that takes as an argument *N* and returns vectors *t* and *y* so that `plot(t,y)` will plot *N* points of the above waveform from $0 \leq t \leq 1$. An outline of the function, where additional code statements are needed, is

```
import numpy as np
def labProblem5_14(N):
    t = np.linspace(0, 1, N)
    y = np.zeros(N) # preallocate vector array
    # statements
    return t, y
```

b) Create a plot of the waveform for $0 \leq t \leq 1$ using 100 points and use a thick line width of 3 to make the line legible against the axes.

5.15 The full-wave nonprecision rectified sinusoid is usually encountered with many cycles, instead of the single cycle computed in Lab Problem 5.14.

a) Create a function that generates `M` duplicates of this waveform. This function should take as arguments `N` and `M`, where `N` is the number of points per cycle and `M` is the number of cycles. It should return vectors `t` and `y`, so that `plot(t,y)` draws the `M` cycles. Hint: It should call `labProblem5_14()`. The function definition is

```
import numpy as np
def labProblem5_15(N, M):
    # statements
    return t, y
```

b) Create a plot of 10 cycles of the waveform using 100 points per cycle. Plot the result using a thick line width of 2 to clearly show the plot against the axes.

5.16* A Collatz sequence refers to a sequence that begins with any positive integer a_n, and whose next value a_{n+1} is computed recursively as half of a_n if that results in an integer, or else three times a_n plus one. Mathematically stated, a Collatz sequence appears as

$$a_{n+1} = \begin{cases} \dfrac{a_n}{2}, & a_n \text{ even} \\ 3a_n + 1, & a_n \text{ odd} \end{cases} \tag{5.12}$$

For example, if $a_1=7$, the sequence is 7, 22, 11, 34, 17, 52, 26, 13, 40, 20, 10, 5, 16, 8, 4, 2, 1. The Collatz conjecture is a famous unproven mathematical problem that says that the sequence will always reach 1 for any positive integer start value. Write a function that takes an integer a and returns the vector holding the Collatz sequence v that begins with a and ends with 1. The function definition is

```
def labProblem5_16(a):
    # statements
    return v
```

Hint: Since the length of the result is not known at the start, the vector v must grow with every iteration. Consider whether a for loop or a while loop is the more natural iteration structure.

5.17 Use the result of the previous problem to write a function that determines which starting number, from 1 to a given integer N, has the longest Collatz sequence. The function should return two numbers, the starting integer and the length of the resulting sequence. For example, if $N = 5$, then it should return the integers 3 and 8, since the Collatz sequence starting with 3 is 3, 10, 5, 16, 8, 4, 2, 1, and at length 8, it is longer than the Collatz sequences starting with any other number from 1 to 5. Hint: This function should call `labProblem5_16()`. The function definition is of the form

```
def labProblem5_17(N):
    # statements
    return longestA, longestLength
```

5.18 Write a function that models the amount of money saved by an engineer. The engineer earns E dollars at the start of every month and automatically deposits P percent of that into a saving account. The savings account earns I percent interest monthly, which is deposited at the end of the month.

Write a function that returns how much is in the savings account at the end of the month after M months. For example, if the engineer has a monthly salary of $10,000 and saves 10% of that in a saving account earning 1% monthly, $E =$ 10,000, $P = 10$, and $I = 1$, after the first month she will have saved 10% of $10,000, which is $1000, and earned 1% interest on it for a total of $1010. After the second month she will have deposited an additional $1000 for $2010 total and then earned a 1% interest on that for a total of $2010 + 20.1 = 2030.10. The function definition is

```
def labProblem5_18(E, P, I, M):
    # statements
    return saved
```

To test it, using the reasoning above, `labProblem5_18(10000,10,1,2)` should return 2030.10, which is 2.030e+03 in scientific notation.

5.19 Modify the problem above so that at the end of every year, the engineering receives a raise of R percent. Hint: Inside the monthly for loop, at the end, check to see if the month is evenly divisible by 12, and if so, modify her salary. The function should be of the form

```
def problem5_19(E, P, I, M, R):
    # statements
    return saved
```

To test it, after 10 years, or 120 months, of saving 10% with a starting monthly salary of $10,000 in an account that earns 1% monthly, and earning a 5% annual raise, `problem5_19(10000, 10, 1, 120, 5)` should return $278,694.23, or 2.7869e+05 in scientific notation. Format your answer to two decimal places.

5.20 A square wave is applied to the circuit shown in Fig. 5.49 that has a capacitor value of $1/k$ F.

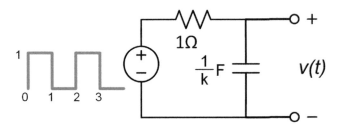

Fig. 5.49 Square wave applied to a circuit (Lab Problem 5.20).

The following piecewise-defined voltage $v(t)$ is measured, where $V_0 = \frac{1}{1 + e^k}$:

$$v(t) = \begin{cases} 1 + (V_0 - 1)e^{-kt}, & 0 \le t < 1 \\ (1 - V_0)e^{-k(t-1)}, & 1 \le t < 2 \\ 1 + (V_0 - 1)e^{-k(t-2)}, & 2 \le t < 3 \\ (1 - V_0)e^{-k(t-3)}, & 3 \le t \le 4 \end{cases} \qquad (5.13)$$

a) Use looping to create a function that takes a value of k and a number of points N in the time region $0 \le t \le 4$ and returns the corresponding time vector t and voltage vector v. The first lines are

```
import numpy as np
def labProblem5_20(k,N):
    t = np.linspace(0, 4, N)
    # statements
    return t, v
```

b) Plot the result for $k = 4$ using many points to make the graph smooth (e.g., $N = 1000$). This is the typical "ocean-wave" waveform encountered after low-pass filtering a square wave.

c) Repeat the calculations using $k = 1/4$. This demonstrates how a simple circuit with a resistor and capacitor can turn a square wave into a triangle wave.

5.21* Many engineering systems can be modeled as being in different states that change with fixed probabilities, called a Markov chain. An example is shown in the diagram in Fig. 5.50, which models how a student who matriculates as an EE major might change his major over four years of education. Here the probability of changing major occurs once at the end of each non graduating year, for a total of three transitions (rising sophomore, rising junior, and rising senior).

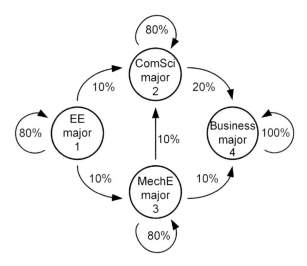

Fig. 5.50 Markov chain of students changing major (Lab Problem 5.21).

a) Write a function called markovMajor() that takes a state, coded as 1 for EE, 2 for CS, 3 for ME, and 4 for business, and returns a state calculated at random according to the probabilities in the Illustrated Markov chain. The function definition is of the form

```
def markovMajor(inCode):
    # statements
    return outCode
```

Test it by verifying that, for instance markovMajor(4) always returns 4, but markovMajor(1) usually returns a 1, but occasionally returns a 2 or 3.

b) Write a program called graduation that determines the graduating major for a student that starts as an EE major (i.e., after going through three potential state changes). It should take no arguments and return a number from 1 to 4. The program should call markovMajor inside it. Hint: The program will be very short since most of the work is done by markovMajor. The function definition is of the form

```
def graduation():
    # statements
    return major
```

c) Write a program that plots the histogram of graduating majors for 100,000 entering EE students. It should call `graduation()`, which will make the resulting code relatively short compared with `markovMajor()`. Include the code and the resulting histogram of output states. The first few lines of the function definition are

```
import numpy as np
def problem5_21():
    N = 100000
    major = np.zeros(N)
    # statements
    return None
```

5.22 Recalculate parts a-c of Lab Problem 5.21 using the Markov chain shown in Fig. 5.51.

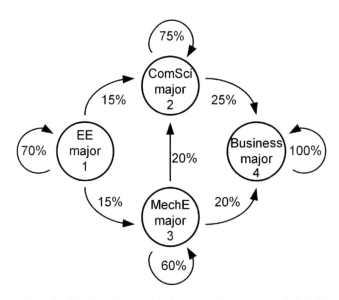

Fig. 5.51 Recalculated Markov chain model of students changing major (Lab Problem 5.22).

5.23 A common method of analyzing fault-tolerant, self-repairing systems uses Markov chains, an example of which is shown in Fig. 5.52. This example shows a system that can tolerate a failure in either Part A or Part B but not both. A single failure may be repaired, but a failure in both parts results in a nonrecoverable system failure. Every day a new random transition takes place.

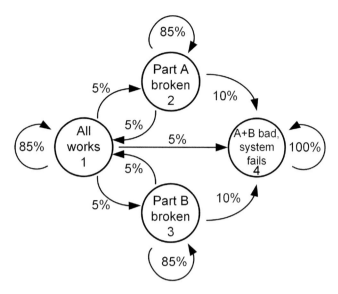

Fig. 5.52 Markov chain model of a fault-tolerant system (Lab Problem 5.23).

a) Write a function called `markovSystem()` that takes a state, coded as 1 for all parts working, 2 representing the case in which Part A is broken, 3 representing the case in which Part B is broken, and 4 representing both parts broken, with resulting permanent system failure. The function should return a state calculated at random according to the probabilities in the illustrated Markov chain. The function definition is

```
def markovSystem(inState):
    # statements
    return outState
```

Test it by verifying that, for instance, `markovSystem(4)` always returns 4, but `markovSystem(2)` usually returns 2, but occasionally returns 1 or 4.

b) Write a program called `week` that determines the state that the computer will be in after 1 week (i.e., after going through six potential state changes). It should take no arguments and return a number from 1 to 4. The program should call `markovSystem()` inside it. Hint: It will be very short since t most of the work is done by `markovSystem()`. The function definition is

```
def week():
    # statements
    return state
```

c) Write a program that plots the histogram of states that the initially working system will be in after a week over 100,000 different random simulations. It should call `week()`, which will make the resulting code relatively short compared with `markovSystem()`. Include the code and the resulting histogram of output state. The first few lines of the function definition are

```
import numpy as np
def labProblem5_23():
    N = 100000
    state = np.zeros(N)
    # statements
    return states
```

5.24 Recalculate parts a–c of Lab Problem 5.23 using the Markov chain shown in Fig. 5.53, which models a system with higher failure rates but improved repair methods.

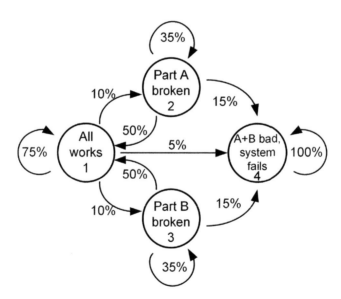

Fig. 5.53 Markov chain system model with higher failure rates (Lab Problem 5.24).

5.25* Filtering, or operating on, data is commonly used in electrical engineering and other disciplines. Stock market analysis, for example, may use a 3-day moving average filter to look for trends while ignoring short-term transients. A 3-day moving average takes a vector of daily stock prices and returns a vector of the same length, representing the same time span, whose corresponding values are the average of three days' worth of current and previous data, where possible.

For example, the output vector's value at index 20 is the average of the input vector's value at 18, 19, and 20.

There are two cases where this is impossible to do. Neither of the first two values in the output vector can be calculated this way because there does not exist enough data in the input vector to do so. The second value of the output vector can only be the average of the first two values of the input vector since one cannot go back an additional index in the input vector. It does not exist. Similarly, the first value of the output vector is the first value of the input vector because earlier values in the input vector do not exist. For example, the input might list Monday–Friday's stock prices as [46 48 50 61 69]. The output for a 3-day moving average filter is [46 47 48 53 60], since 60 in the output vector is the average of 50, 61, and 69 in the input vector, and in the output vector, 47 is the average of 46 and 48, as there is no input value prior to 46.

Write a function that takes an input vector of data of arbitrary length and returns a vector of the same length that is the 3-day moving average of the input. The function should be of the form

```
import numpy as np
def labProblem5_25(vIn):
    N = len(vIn)
    vOut = np.zeros(N) # preallocate vector NumPy array
    # statements
    return vOut
```

5.26 Generalize the function that solves Lab Problem 5.25 so that the function is a moving average filter of length *M*, where *M* = 3 in the previous problem. Hint: The NumPy mean() command calculates the average of a vector array. A few lines of the function are

```
import numpy as np
def labProblem5_26(vIn, M):
    N = len(vIn)
    vOut = np.zeros(N) # preallocate vector NumPy array
    # statements
    return vOut
```

Test it using the fact that

```
import numpy as np
labProblem5_26(np.array([46, 48, 50]), 2)
```

returns the `NumPy array [46., 47., 49.]`.

5.27 The built-in function `exp(x)` computes e^x, but it could also be computed in the limit as *N* grows large using the formula

$$e^x = 1 + \sum_{k=1}^{N} \frac{x^n}{n!} \qquad (5.14)$$

where *n*! is the product $1 \times 2 \times 3 \times \cdots \times n$. Create a function that calculates this based on the function definition

```
def labProblem5_27(x, N):
    # statements
    return result
```

Test it by confirming that `labProblem5_27(1,100)` closely approximates $e^1 = e = 2.718$. Hint: There is a predefined `math module` function called `math.factorial()`.

5.28 The sum

$$f(N) = \sum_{k=0}^{N} \frac{\sqrt{12}}{(-3)^k (2k+1)} \qquad (5.15)$$

will calculate π in the limit as `N` grows to infinity. Calculate this sum by writing a function of the form

```
def problem5_28(N):
    # statements
    return result
```

To test it, setting `N = 7` should return π to four decimal places of accuracy.

5.29* Use Python to write a function that solves the math problem

```
result = (x + y) * 7
```

The function definition should be of the form

```
def labProblem5_29(x,y):
    # statements
    return result
```

The input arguments will be in base 2 and should be provided with quotation marks. The result should also be in base 2. To check your work, `labProblem5_29('101', '110')` should return `1001101`.

5.30 Write a function that takes a number in decimal form and returns two strings: the string representation of the number in binary and in decimal. The function definition should be of the form

```
def labProblem5_30(decimalInput):
    # statements
    return binStr, decStr
```

To test it, note that `labProblem5_31(11)` should return

```
binStr = '1011'
decStr = '11'
```

5.31 Fourier series analysis is an important technique from the signal processing subdiscipline of electrical engineering. It involves the fact that any periodic waveform can be broken up into a sum of sinusoids. It sounds impossible that sharp-edged function such as a square wave could be composed of a sum of smooth sinusoids. Indeed, Fourier's original 1807 paper on the subject was denied publication for this reason! In the limit of summing an infinite number of sinusoids, however, it is true.

As an example, create a Python function of the form

```
def labProblem5_31(N):
    # statements
    return f
```

that calculates the function

$$f = \sum_{k=1,3,5,\ldots}^{N} \frac{\sin(kt)}{k} \tag{5.16}$$

Use a 1000-element time vector array for t ranging from $0 \leq t \leq 20$. In words, first create a time vector array t, and then initialize the result vector f as a vector array of zeros that is the same size as the t vector array. Next, create a for loop using a loop variable k that starts at 1 and increments in steps of 2 so it becomes every odd number up to N. For every iteration of the loop, add the vector `sin(kt)/k` to `f`. This is a vector array because `t` is a vector array. The function returns `f`, but check your work by embedding a `plot(t,f)` at the end of your code. Test it with a value of `N=49`. It should look similar to a square wave. As the upper limit of the sum, 49, is increased, the function will appear progressively more similar to an ideal square wave. When the program is run, besides returning the vector `f`, it should also display the plot. Place this plot in a Word document.

5.32 Challenging! You want to create a 72 Ω resistor, but that is not a standard value. One way to build it is with 33 Ω and 39 Ω resistors in series. Another way is with 75 Ω and 1800 Ω resistors in parallel. Numerically, both these options work out to the same 72 Ω resistor. You need to evaluate how resistor tolerances affect the equivalent resistance in both cases. If you do a Monte Carlo analysis of the two options with 10,000 simulations, do the two options have the same spread of values if they are all made with 5% resistors? For this problem, the function should create the two histograms of data, labeled to show which is the series and which is the parallel solution. These plots should be placed in a Word document with additional text indicating your evaluation of which is the better method or if the two are equally preferable. The `NumPy mean()` and `std()` functions can be used on your data arrays to support your conclusion.

SPICE

6

6.1 OBJECTIVES

After completing this chapter, you will be able to use LTspice to do the following:

- Create a schematic
- Analyze a DC circuit and find the internal voltages and currents, as if using a DMM
- Analyze a time-varying circuit and find internal voltages, as if using an oscilloscope
- Analyze a circuit whose output changes with frequency of the input
- Describe the several meanings of "ground"
- Describe the difference between energy and power, and use LTspice to compute them
- Simulate two common types of ICs: the 555 timer and the operational amplifier

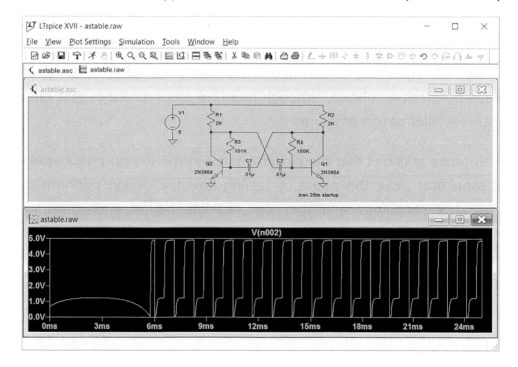

6.2 CIRCUIT SIMULATION

In previous chapters, many of the programming exercises included calculations related to circuits. As the circuits get more complicated, it is often helpful to be able to simulate their function, especially if the input signal is not a constant. This chapter will introduce you to an often-used simulation environment called SPICE. General circuit simulators calculate the voltages and currents in arbitrary circuits. They have been in existence since the early 1960s, encouraged by increasingly complex circuit designs and initially funded by the US Department of Defense. An undergraduate class project at Berkely spurred some of the first precursors to the open-source program we now know as SPICE, the Simulation Program with Integrated Circuit Emphasis. The first complete version was authored by Laurence Nagel in 1973. Since then, SPICE has evolved to become the de facto open-source standard for all circuit simulation packages.

Most popular modern circuit simulators use SPICE at their core, including LTspice by Analog Devices, PSPICE by OrCAD, and Multisim by National Instruments. This text uses LTspice, which is free and available for Windows and MacOS; it also works well under the Wine compatibility layer for Linux.

6.3 UTILITY

There are often two stages involved in solving real-world engineering problems: **design** and **validation**. In the design phase, the engineer uses the problem specification to create a schematic using techniques taught in circuit and logic design courses. The design phase is not generally aided by circuit simulation, although there are other packages that can assist with this phase, such as filter design programs.

The validation phase confirms that the circuit created in the design phase works as intended. Circuit simulators assist this phase by testing how the design performs under both ideal and nonideal conditions, and they can evaluate performance given a range of component tolerances, temperatures, and operating conditions. This chapter introduces the circuit simulator LTspice and provides examples of using it, with Python, to assist the validation phase.

6.4 INSTALLING LTSPICE

Download LTspice from the Analog Devices website (http://www.analog.com). Versions are available for both the PC and MacOS. Both are relatively small downloads, and despite possible differences in version numbers, both offer near-identical user experiences. Microsoft users should choose the 64-bit version if prompted during installation. When the installation is complete, the program will automatically load and show the screen in Fig. 6.1.

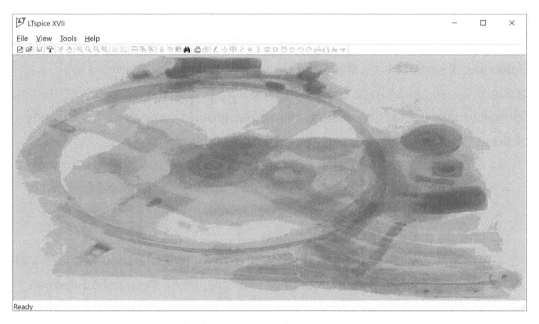

Fig. 6.1 LTspice start screen. The LTspice application allows you to create, open, test, and save circuits.

6.5 STEPS IN SIMULATING

This textbook describes the following three steps in circuit simulation in detail:

1. Draw the schematic, that is, schematic capture. The following different types of entities must be defined in this stage:
   a. Components, such as voltage and current sources, resistors, and capacitors.
   b. Ground.
   c. Wires connecting the above elements.
2. Define the analysis type. In this textbook, we will describe the three most common types:
   a. DC analysis: DC analysis assumes all sources are constant, so the outputs will be constant.

b. Transient analysis: Transient analysis measures time-varying outputs.

c. AC analysis: AC analysis measures how the circuit responds to different frequencies of a sinusoidal input source.

3. View the results. Depending on the analysis type, this output may be a number, such as a reading from a digital voltmeter; a time-dependent waveform, such as an oscilloscope trace; or a frequency-dependent Bode plot.

TECH TIP: WHAT IS GROUND?

Unlike most components, there are three commonly used symbols for ground: $\equiv \; \underline{\vee} \; \not\!\!/$. LTSpice uses the triangular symbol shown in the center. To make matters more confusing, any of the ground symbols can mean any of three different concepts, depending on the context:

1. Ground can mean a physical connection to the earth, often through a copper spike driven into the ground. Grounds are required by certain types of radio antennas, as well as by electrical power safety codes, to protect the user in case insulation fails.

2. Ground can also refer to the set of conductors that return current to the power source in a circuit. For example, most electrical equipment in automobiles have two power connections: one connected by wires through a fuse to the +12V source and the other to the metal chassis of the car, called "ground". The car chassis is connected to the negative side of the electrical power source by a thick ground wire.

3. In this chapter, however, ground refers to an imaginary, arbitrary voltage reference point in a circuit. Voltage is always measured between two points, which makes it awkward to describe in a circuit simulator, since the number of possible combinations of any two points in a circuit rises very rapidly with circuit complexity. When the user attaches the ground symbol to a point, say at "Point B", she is saying all voltages are measured relative to Point B. A single ground symbol does not change the circuit behavior in any way; it just lets us say "the voltage at Point A is 6 V", rather than "the voltage between Point A and Point B is 6 V". In other words, we define the voltage at

ground to be 0 V, which lets us measure all other circuit voltages relative to ground.

Ground is often placed at the most negative point in the circuit, which is often the negative terminal of the power source. In this case, all other voltages in the circuit (relative to ground) will be positive, and all current in the circuit will return through ground to the power source.

Figure 6.2a shows a circuit with voltages defined between various points. The adjacent schematic shown in Fig. 6.2b adds a ground symbol, which lets the same voltage be redefined in a much simpler manner. Which would you rather read?

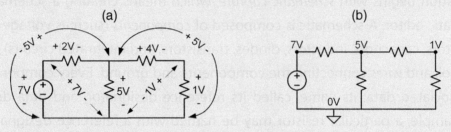

(a)   (b)

Fig. 6.2 Voltages in circuits with different references. (a) Voltages can be defined between all points. (b) Alternatively, voltages can be references to a single ground.

PRACTICE PROBLEMS

6.1 The schematic shown in Fig. 6.3 shows a circuit with a bipolar junction transistor, or BJT. You may have first seen it in Lab Problem 4.8; however, its operation will be covered in a future electronics course. You do not need to understand how it works to use the concept of ground to simplify the circuit. Voltages between

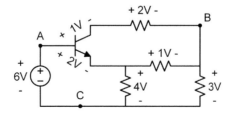

Fig. 6.3 Bipolar junction transistor (BJT) circuit. Using a ground placed at the bottom of the circuit, voltages can be determined at points A, B, and C.

various points are shown. If ground is placed at the bottom of the circuit, determine the voltages relative to ground at points A, B, and C in the circuit. Use the example in the Tech Tip: *What Is Ground* as a guide and remember that voltages do not change along a wire, only across components. Note: This does not use LTspice. Write your answer in your preferred word processor.

6.6 SCHEMATIC EDITOR

Circuit simulation begins with *schematic capture*, which means creating a schematic diagram in the schematic editor. A schematic is composed of *components* (such as voltage and current sources, resistors, capacitors, inductors, diodes, transistors, and integrated circuits), at least one *ground* symbol, and *wires* connecting the components and ground. Every component has two pieces of associated data: its name, called its reference designator, and data describing its value. For example, a particular resistor may be named with a reference designator R11 and have a value of 4.3 kΩ, or a voltage source may be called V1 and have a value of 12 V.

6.6.1 Starting a New Schematic

The Windows application start button appears as [LTspice XVII Desktop app]. When started, LTspice periodically checks for new devices added to its library. Analog Devices provides this circuit simulator for free because its tight integration with the company's integrated circuits en-courages engineers to use its devices. Since this textbook uses only basic components, it is safe to skip these library updates if you wish.

Once loaded, the program interface will appear as shown in Fig. 6.4. The grayed-out field, in the center of the screen, indicates that there is no current schematic. To create one, choose File → New Schematic from the menu. The logo will disappear, and the program's title bar will show "Draft1" to indicate that a new schematic has been created with the default name. Change the name to "Schematic1" by using File → Save As and notice the change in the title bar text.

Fig. 6.4 Starting a new schematic. The name of the new schematic is shown circled in *red* in the *upper left*.

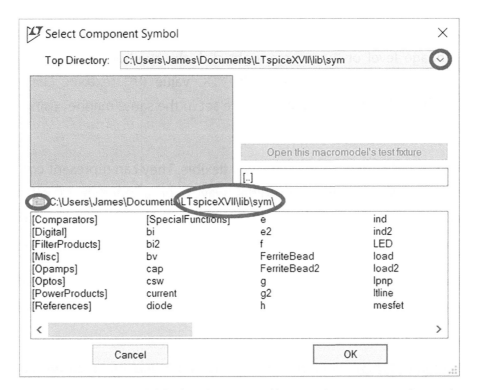

Fig. 6.5 Select component symbol dialog box. The component library provides access to many electronic devices.

6.6.2 Selecting the Library

The library is a collection of predefined components, including voltage and current sources, resistors, inductors, capacitors, and integrated circuits, all of which can be selected and drag-and-dropped to create a schematic. All these components, and the libraries they are contained in, are selected using the ⊅ button on the main toolbar. Choose it to display the Select Component Symbol dialog box, as shown in Fig. 6.5.

Note the folder in which the library is defined, highlighted with a red ellipse in Fig. 6.5. If the component dialog does not show components and component folders beginning with the [Comparators] folder, it is because the library folder is pointing to the wrong location. Change it to the C:\Users\UserName\Documents\LTspiceXVII\lib\sym folder using the Top Directory drop-down selector and the Move Up Folder selector highlighted in Fig. 6.5.

6.6.3 Defining Voltage and Current Sources

To create a voltage source, select the component option using the ⚡ button on the main toolbar. The component dialog box should appear as shown in Fig. 6.5, still set to the correct library location selected in the above step. Move the horizontal scrollbar until "voltage" is highlighted. Select it, click to place it in the schematic, and press the Esc key to escape from voltage source placing mode.

To select the voltage level of the source, right-click the symbol, not the text next to the symbol, and enter the desired voltage in the "DC value (V)" textbox. Leave the "Series Resistance (Ω)" textbox blank. Current sources are set in the same manner using the "current" component.

The LTspice voltage and current sources are very flexible. They can represent constant voltage or current sources, square waves, sinusoidal waves, or a number of other commonly encountered waveform types. By default, they are constant DC sources. To select a different waveform type, right-click the symbol, not the text, and choose the "Advanced" button.

To change the designator ID of the source, right-click the text surrounding the source. Typical names for voltage sources are V1, V2, and the like, and I1, I2, and the like for current sources.

6.6.4 Other Common Components: Resistors, Capacitors, Inductors, Diodes, and Transistors

Resistors, capacitors, inductors, diodes, LEDs, transistors, and more can be selected using the component dialog box. The only abbreviated names are "res," "cap," and "ind" for resistor, capacitor, and inductor. These three and the diode also have dedicated toolbar buttons ⚡ ⚡ ⚡ ⚡. You can use the prefix symbols M, k, m, and u to stand for Mega, kilo, milli, and micro, respectively. Note that the letter u is used instead of μ. For example, it is less error-prone to specify a 2.2 μF capacitor as 2.2u rather than as 0.0000022.

6.6.5 Manipulating Component Placement

Components can be rotated in 90-degree increments using the Ctrl+R key combination before they are placed. You can also reflect the component about the vertical axis using the Ctrl+E combination. Once they are placed, they can be rotated, reflected, or moved by selecting the move tool, ✋, selecting the component, and then either moving it with the mouse, pressing Ctrl+R, or pressing Ctrl+E to reflect it. Components can be deleted by selecting the ✂ tool and then clicking on the component to be deleted. Mistakes can be undone with the undo tool ↺; however, the undo tool is not mapped to the typical Ctrl+Z shortcut.

6.6.6 Wire and Ground

Draw wires by selecting the wire tool, ✐. Unwired connection points on components appear as open squares. Once connected with a wire, the square disappears. The ground symbol is placed after selecting the ground tool, ⏚. Make certain the open square on top of the ground disappears after it is connected to the rest of the schematic with a wire. If this is not done, or if the ground is not connected, the simulation with fail and report the warning: "This circuit does not have a conduction path to ground."

6.6.7 Operational Amplifiers and other Integrated Circuits

Future courses will likely introduce a myriad of new components, called integrated circuits (the "IC" in SPICE). These can be found from the component dialog box by opening the folder of the IC category, and they include components such as comparators and opamps. Sometimes generic components will be available; an example of this is the "UniversalOpamp2" available in the "Opamps" library, which is meant to represent a generic, somewhat idealized opamp.

6.6.8 Schematic Capture Summary

- Schematic capture is the first step of circuit simulation.
- Components: created using ᗇ toolbar button.
  - Examples: voltage sources, resistors, inductors, transistors, ICs.
  - Rotate them using Ctrl+R and reflect them using Ctrl+E before placing.

- Reference designator: right-click on the text to change the component name (e.g., C2).
- Value: right-click on the symbol to change the value, for example, 4.7 μF.
- Ground: created using the ⏚ toolbar button.
  - At least one required per circuit.
  - Often placed at most negative part of the circuit.
- Wires: created using the ✐ toolbar button.
  - Open squares at connection points disappear when connected with a wire.

6.2 Create a schematic of the circuit shown in Fig. 6.6, print it, and annotate it with your name. Save it to your work folder to be used in the next Practice Problem. Using the Windows Snip & Sketch Tool or the Mac Grab Tool, cut and paste the image into your preferred word processing program. ®

6.3 Write a Python function with input parameters Vin, R1, and R2 that returns the voltage drop across R1 and the voltage drop across R2. The function is partially written below; fill in the sections marked with "#". The default values of the input parameters are set to those shown in Fig. 6.6. Your function should have the form

```
>>> def voltage_divider(Vin=12,R1=2,R2=4):
...      # Body of function to calculate
...      # V1 or Voltage across R1
...      # V2 or Voltage across R2
...      return VR1, VR2
...
>>> voltage_divider()
```

Check that your program gives the theoretical values described in the Chapter 2 Tech Tip *Voltage Dividers* at the end of Section 2.13.

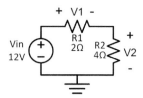

Fig. 6.6 Schematic capture of a voltage divider circuit.

6.7 TYPES OF LTSPICE SIMULATIONS

There are three commonly used SPICE analysis types at the undergraduate level: DC operating point, transient, and AC sweep.

6.7.1 DC Operating Point

This is a complicated name for a simple concept. Use this analysis type if the circuit has only unchanging sources, meaning constant-voltage and constant-current sources. In this case, all voltages and current in the circuit will also be constant and would be measured in a real-world circuit using a digital multimeter (DMM). There are no options for this simulation, and the results are viewed by hovering the cursor over the wire or component of interest.

6.7.2 Transient Analysis

Transient analysis is used if the circuit has changing voltages and currents. Examples include circuits whose sources step from one voltage to another, or are defined by a square wave, and circuits that oscillate. The real-world circuit would require an oscilloscope to view how the circuit voltage and current changed in time. The simulated results are viewed by clicking the wire or component of interest after simulation to show their voltage on a simulated oscilloscope.

6.7.3 AC Sweep

Electrical engineers are often concerned with how a circuit output changes when the input is a purely sinusoidal source of varying frequencies. For instance, a high-fidelity audio amplifier should exhibit the same signal amplification, or gain, for input frequencies spanning the range of human hearing, often taken to be 20 Hz—20 kHz. Another example is a low-pass filter placed before a bass speaker amplifier that passes frequencies below about 150 Hz, but attenuates higher-frequency signals. This analysis is performed with an alternating current (AC) sweep, in which one source is defined to be the source of the variable-frequency AC signal, and the resulting changes in the signal magnitude and phase are displayed versus frequency. It is called AC because the direction of the current from the source constantly alternates, unlike the direct current (DC) sources we see used, for instance in Fig. 6.7, that maintain a constant direction of current.

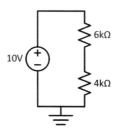

Fig. 6.7 LTspice DC circuit simulation of a voltage divider circuit. Using LTspice, you will find the DC circuit voltages at different points in the circuit.

This analysis is the equivalent of attaching a spectrum analyzer to a real-world circuit, and the resulting magnitude versus frequency plot is commonly called a Bode plot. Bode plots were described in the Chapter 3 Tech Tip *Bode Plots* following Section 3.2.9.

6.8 DC SIMULATION WALKTHROUGH

Once the first phase of circuit simulation, schematic capture, is completed, the second and third phases, defining the analysis type and viewing the results, must be accomplished. These are best taught by example. The first of the three main types of circuit analysis this chapter covers is DC simulation, and it will be taught by completing an analysis of the circuit shown in Fig. 6.7.

PRACTICE PROBLEMS

6.4 Use your Python program developed in Practice Problem 6.3 to find the voltages across the two resistors shown in Fig. 6.7. In this case you will need to override the default values in your original program. ®

6.8.1 Open a Blank Schematic

- Open LTspice and from the menu, File → New Schematic.
- Save it immediately using the menu, File → Save As. Give the schematic a simple name with no spaces under "Name" (e.g., spice_dc). Save it in your personal data directory for this course.

6.8.2 Add a 10 V Source

- Click the component icon ⊅. Find the "voltage" component. Typing the first few letters will scroll to it. Double-click to select it and place it in your schematic by clicking once. If you click again, you will place a second voltage source. To prevent this and get out of "create voltage source" mode, press the Esc key.
- To delete and move components, rather than select and drag or select and delete, as in most programs, choose the action first, and then select the thing you want to change. Create a second voltage source now using the above steps to practice.
  - To delete the second voltage source, select the delete tool ✂, and then click on the voltage source you wish to delete. Press Esc to exit delete mode.
  - Move your desired V1 source to the left middle screen by selecting the move tool ✋. If you forget which icon it is, hover the mouse over the icons for a tooltip. Select the voltage source you want to move, and then move your mouse and click to place it. Press Esc to exit move mode.
- To set the voltage to 10 V, right-click inside the symbol. Do not right-click on text outside the symbol or you will change the *name* of the source, that is, its reference ID, but not its *value*.
  - Set the DC value to 10 V. The "V" at the end is optional. Leave the series resistance box empty and press OK.
- Save your work: Ctrl+S or from the menu, File → Save.
- Review: To make a 10 V source, select the part using ⊅ and choose "Voltage", and then set the value of the source by right-clicking inside the symbol.

6.8.3 Add Two Resistors

- You could click ⊅ and choose "resistor", but resistors are so common they have their own icon, ⧓. Click it and place two resistors to the right of the voltage source, stacked vertically but separated by a small distance. Click "Esc" to exit resistor-placing mode.

- Test your ability to rotate components by placing one resistor horizontally oriented. To do this, choose the resistor icon again as you did above, but before placing the resistor, press Ctrl+R to rotate it. Place this horizontal resistor, and then delete it using the technique you learned in the previous section.
- Right-click the resistor (not the text) and make the top one 6 k and the bottom 4 k. Save your work.
- Review: To place a resistor, choose ⟨, and right-click the symbol to set the resistor's value. Press Ctrl+R before placing to rotate.

6.8.4 Add the Ground and Wire Everything Together

- Click the ground icon ⏚ and place it below the voltage source so there is a little space separating them.
- Select the wire tool ✎ to wire the parts together.
- Click on one component end, and then click on another component that you wish to wire together. The open square ends of the wire will vanish, and the wire will stop following the cursor. Clicking intermediate points while wiring will allow you to create corners.
- Mistakes can be deleted using the ✂ tool.
- Repeat to wire all components together. Compare your result with Fig. 6.8 and save your work. Hint: To check your work, check to make sure there are no open boxes, that is, unwired connection points.
- Review: Place a ground using ⏚ and wire components using ✎.

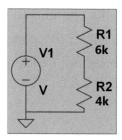

Fig. 6.8 DC circuit simulation.

6.8.5 Run the Analysis

- Specify the type of analysis to run. From the menu, choose Simulate → Edit Simulation Cmd and choose the "DC op pnt" tab (DC operating point) to determine all voltage and currents for this example with constant sources. Choose "OK" and place the ".op" text that appears anywhere on a blank area of the schematic.
- Select Simulate → Run from the window or choose the run button ⚡. An analysis window will appear and show the voltages at, and currents through, various cryptically named nets connected by wires such as "n001".
- Measuring and labeling voltage: To determine the voltage along any wire relative to ground, hover the cursor above the wire and look in the status bar at the bottom of the window. This simulates probing the wire with a digital multimeter. Click any wire to place a label with its voltage on the schematic.
- Measuring current and power: To determine the current flowing through any component and the power that it dissipates, hover the cursor over the component and look in the status bar.
- Review: Find DC voltages and current using Simulate → Edit Simulation Cmd and choosing "DC op pnt". Run the simulation using the ⚡ button. View voltages and currents by hovering over them in the schematic window. Label voltages on the schematic by left-clicking a wire.

PRACTICE PROBLEMS

6.5   Return to the schematic you first captured in Practice Problem 6.2 and that is shown in Fig. 6.9, and now that you know how to analyze it in LTspice, find and label the voltage across the 4 Ω resistor. ℝ

6.6   Compare your results obtained above in Practice Problem 6.5 with those from your Python program in Practice Problem 6.3. Do your results agree?

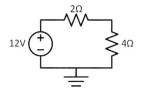

Fig. 6.9 Using LTspice to find the voltage across the 4 Ω resistor.

Many people use the terms "power" and "energy" interchangeably, but these terms have very different meanings. Energy, measured in joules (J), is the ability to do work — examples of a quantity of energy include a coin lifted by 1 foot (20 J), two AA batteries (20 kJ), a ham sandwich (2 MJ), or 20 gallons of gasoline (2.4 GJ). The rate at which energy is used is power, measured in watts (W, or J/S). A Ferrari will be able to deliver more power than a Volkswagen Beetle, even though both may start with their gas tanks identically full. That is, they begin with the same energy, but the greater power of the Ferrari will cause it to use that energy more quickly. A car going up a hill will use more power than the same car traveling on flat ground. The ham sandwich will power a young child for longer than an adult, and the AA battery will light a 150 mW LED for about twice as long as a brighter LED rated for 300 mW of power consumption.

Mathematically, energy is the time-integral of power:

$$\text{energy} = \int \text{power}(t)dt \qquad (6.1)$$

If the power is constant, this simplifies to

$$\text{energy} = \text{power} \times \text{time} \qquad (6.2)$$

Power is the time derivative of energy,

$$\text{power} = \frac{d}{dt}\text{energy}(t) \qquad (6.3)$$

or, if the power is constant, this simplifies to

$$\text{power} = \frac{\text{energy}}{\text{time}} \qquad (6.4)$$

6.7 Use LTspice to determine how much power is being delivered by the voltage source in Fig. 6.10. ®

6.8 Continue to analyze the circuit shown in Fig. 6.10, first introduced in earlier Practice Problems. How long will eight 1.5 V AA batteries power the circuit if a single AA battery holds 10 kJ of energy? Hint: Solve the power = energy/time Eq. 6.4 given in the *Power versus Energy* Tech Tip. ©®

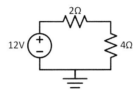

Fig. 6.10 Using LTspice to find the power delivered to the circuit.

PRO TIP: ETHICS II

Your engineering integrity will likely be tested many times during your career, which is why the Accreditation Board for Engineering and Technology (ABET) requires that ethics be an integral part of engineering education. Besides the IEEE Code of Ethics, described in an earlier Pro Tip, engineers who earn the professional engineering (PE) certification agree to follow a similar set of rules set by the National Society of Professional Engineers (NSPE). The preamble to the NSPE Code of Conduct for Engineering (2007) states:

Engineers shall at all times recognize that their primary obligation is to protect the safety, health, property, and welfare of the public. If their professional judgement is overruled under circumstances where the safety, health, property, or welfare of the public are endangered, they shall notify their employer or client and such other authority as may be appropriate.

David Jonassen developed a generic framework for applying the NSPE Code of Ethics, or the IEEE Code of Ethics, to any engineering ethics problem

(Journal of Education, vol. 98, no. 3, 2009). His analysis method consists of five steps:

- State the problem. Define the ethical problem clearly in a few words.
- Get the facts.
- Identify and defend competing moral viewpoints. Most ethical problems involve multiple perspectives, and each viewpoint may lead to a different conclusion.
- Identify the best course of action.
- Qualify the course of action with facts.

When faced with a difficult decision, consider documenting your decision using this framework.

TECH TIP: 555 TIMER

There are over ½ million integrated circuits (ICs) sold just by Digi-Key, one of the largest electrical parts vendors. Of these, the 555 timer is historically the most popular. Since its introduction by Signetics in 1972, this chip has outsold all other types of ICs. It has been used in medical monitors, spacecraft navigation systems, and children's toys. Although it can be used in many different configurations, one of the most common is as an astable multivibrator — in simple words, an oscillator. The frequency is set using two resistors and a capacitor; it does not change significantly with change in the powering voltage. The schematic using it as an astable multivibrator and the pin numbering for the IC are shown in Fig. 6.11.

Notice that although the schematic symbol is rectangular, like the IC, the pin numbering of the schematic does not have to correspond to the pin positions on the actual device. This allows the schematic to be drawn more simply. For instance, the schematic shows pins 4 and 8 next to each other, since they are electrically connected with a wire, although on the physical device, those pins are diagonally across the body of the IC.

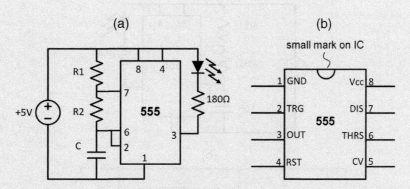

Fig. 6.11 NE555 circuit schematic and pinout diagrams. (a) The NE555 circuit diagram shows a simple connection to external resistors, capacitors, a voltage source, and an LED. (b) The pinout diagram shows the pin names of the integrated circuit.

The design equations for this circuit are as follows:

LED blinks on for $0.693 \cdot R2 \cdot C$ seconds

LED blinks off for $0.693 \cdot (R1+R2) \cdot C$ seconds

The duty cycle is the percentage of the time that the output is on. As an example, if it is on for 1 s and off for 3 s, then it is on for 1 s out of every 4 s total, for a duty cycle of 25%.

PRACTICE PROBLEMS

6.9 Use Python to find the values for R1 and R2 such that the LED is on for 1/3 second and off for 2 seconds. Use a 1 μF capacitor. Hint: R1 will be between 1 MΩ and 10 MΩ, and R2 will be between 100 kΩ and 1 MΩ. ®

6.9 TRANSIENT SIMULATION

The second of the three types of circuit analysis this chapter describes is transient analysis, where circuits are analyzed with varying voltages or currents. It will be demonstrated by completing an analysis of the circuit shown in Fig. 6.12a. Note that although the circuit in Fig. 6.12a is the same as that in Fig. 6.11a, this is not the solution to Practice Problem 6.9.

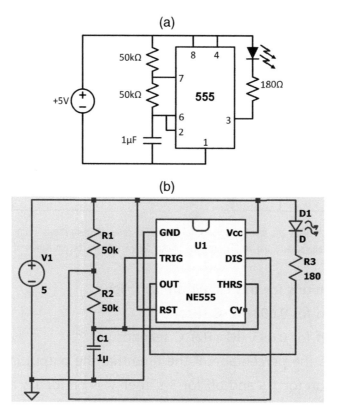

Fig. 6.12 Astable multivibrator circuit schematic and LTspice schematic capture using the NE555. (a) The circuit schematic simplifies the external connections. (b) The LTspice circuit shows a schematic of the actual pin connections to the NE555 integrated circuit.

Follow the steps to gain a solid working knowledge of transient circuit analysis and then apply it to Practice Problem 6.9. Figure 6.12b shows the completed schematic capture. Note that LTspice fixes the location of the IC pins in the same positions as on the physical chip. Although this makes wiring up a physical circuit from an LTspice schematic easier, it makes the schematic more difficult to read. Use both these schematics to understand the application of the NE555 as an astable vibrator.

6.9.1 Open a Blank Schematic

- Open LTspice; from the menu, File → New Schematic.
- Save it immediately using the menu, File → Save As. Give the schematic a simple name with no spaces under "File name", spice_transient, for example. Save it in your personal data directory for this course.

6.9.2 Add a Voltage Source

- Click the component icon ⊅. Find the "voltage" component typing the first few letters will scroll to it — double-click to select it, and place it in your schematic by clicking once. Right-click the symbol, not the text, and make it 5 V.

6.9.3 Add a 555 Integrated Circuit

- Integrated circuits are organized into folders. Click the component icon ⊅, choose the [misc] folder, and select the NE555 component. The NE prefix is unimportant; it indicates a variant improving upon the original 555 design. Leave plenty of room between it and the voltage source to place the resistors and capacitor. You may need to alter the zoom with the ⊕ ⊙ ⊖ ⊠ tools that zoom in, pan, zoom out, and zoom to fit the entire schematic on the visible page, respectively.

6.9.4 Add Resistors, a Capacitor, an LED, and a Ground

- Click the resistor icon ⪦, and Ctrl+R if desired to rotate, place, and right-click the component symbol (not text) and give it a value of 50k. Spice understands "k", and there is no need to add the Ω symbol. Do the same for the capacitor using the symbol ⊥ and using 1u. Recall that LTspice uses "u" for μ. The LED is found under the main ⊅ folder and, like the 555 IC, has no parameters to set. The ground is placed with the ⏚ symbol.

6.9.5 Wire it Up

- Use the ✐ tool to wire the circuit together. When you place a wire on another wire, it will form a closed square indicating a connection, as shown in Fig. 6.12. Look carefully; an open, unfilled square means that end of the wire is not connected to anything else.

6.9.6 Run the Transient Simulation Analysis

- Specify the type of analysis to run. From the menu, choose Simulate → Edit Simulation Cmd and choose the default "transient" tab, since the oscillator's output varies with time. To analyze the circuit from 0 to 0.5 seconds, specify 0.5 seconds as the stop time and 0 seconds as the time to start saving data, and leave the rest of the options blank.

DIGGING DEEPER
Curious about all the other options for transient simulation analysis?
Most of the options are rather arcane and can be investigated under
Help > LTspice > Transient Analysis Options. The only three commonly
adjusted options are *Stop time*, *Time to start saving data*, and *Maximum timestep*. *Stop
time* is the duration of the simulation in seconds. Set a *Time to start saving data* value to
greater than 0 seconds if early circuit behavior is not of interest. LTspice automatically
chooses how frequently to compute output values, more often when output values are
quickly changing and less-frequently otherwise. This can be manually controlled by
setting the *Maximum timestep* value, often in milliseconds or less.

- Choose "OK" and place the ".tran 0.5" text on a blank area of the schematic.
- Select Simulate → Run from the window or choose the 🏃 button. The split
  schematic/analysis window shown in Fig. 6.13 will appear.

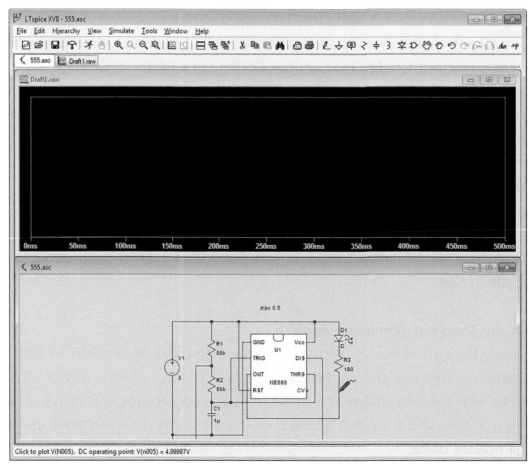

Fig. 6.13 Split schematic and analysis window. The output will be displayed on the top window. On the lower schematic
window, holding the cursor over a wire turns it into a red probe.

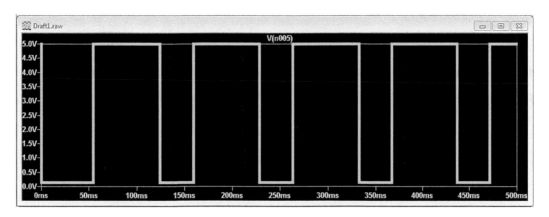

Fig. 6.14 NE555 output waveform. Positioning the red probe over the wire connected to pin 3 and clicking it displays the output voltage vs. time. Clicking a component displays the current through it.

- Position the cursor over a wire in the schematic in which you want to see an oscilloscope-style plot of the voltage versus time. In Fig. 6.13, the cursor is hovering over the wire that is attached to pin 3, OUT. The cursor changes to resemble a red oscilloscope probe.
- Click on the probe to display the corresponding voltage waveform shown in Fig. 6.14.
- Click on other wires to add traces. Right-click on the plot area and from the popup menu choose Edit → Delete; then click on the plot legend at the top of the screen to remove.

PRACTICE PROBLEMS

6.10 Modify the above walkthrough to print 1 second's worth of data of the current through the diode. Record a screen grab of the analysis window using the Snip & Sketch Tool if on Windows Win+Shift+S or Command+Shift+4 if on a MacOS. ®

TECH TIP: OPERATIONAL AMPLIFIERS

The operational amplifier, or more simply "opamp", is a very commonly used integrated circuit. It is available in a stunning number of varieties; Digi-key sells more than 35,000 different types of opamps. Opamps are thoroughly described in circuit courses; this textbook considers one application of an idealized opamp whose schematic symbol is shown in Fig. 6.15.

Fig. 6.15 Operational amplifier symbol.

The opamp is used for many different purposes, including amplifying, filtering, and even as an analog computer, where it integrates and differentiates input signals. An example of a low-pass filter is shown in Fig. 6.16. The signal input $v_{in}(t)$ to the opamp is on the left, the output is on the right, and the vertical wires in the middle of the opamp with the small polarity symbols are the ± 5 V DC power connections to the opamp. The circuit shown in Fig. 6.16, known as a second-order Butterworth low-pass filter, passes input sinusoids with frequencies below a preset frequency f_c, but attenuates sinusoids of higher frequencies. This is similar to the low-pass filter introduced in Lab Problem 5.9. It could be placed before a bass amplifier to ensure that no high frequencies are sent to the woofer speakers. The rightmost 5V source's wire crosses, but does not connect to, the output of the opamp. The convention in EE is that two wires crossing perpendicularly do not connect unless there is a black dot drawn at their intersection.

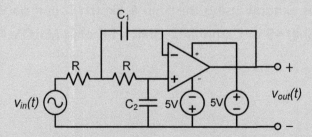

Fig. 6.16 Second-order Butterworth low-pass filter.

The design equations for this circuit, given a desired cutoff frequency f_c and resistor value R, are

$$C_1 = \frac{C_2}{2} \text{ and } C_2 = \frac{1}{\sqrt{2}\pi R f_c} \tag{6.5}$$

PRACTICE PROBLEMS

6.11 Write a program in Python that takes a frequency and resistor value and uses Eq. 6.5 to determine C_1 and C_2. Use it to design a low-pass filter using 10 kΩ resistors with a cutoff frequency of 200 Hz. This is a realistic cutoff value for a subwoofer speaker amplifier. ®

6.10 AC SWEEP SIMULATION

The last of the three types of circuit analysis that this chapter describes is AC sweep analysis, and it will be taught by completing an analysis of the circuit shown in Fig. 6.17a. Note that although the circuits are similar, this analysis does not produce the solution to Practice Problem 6.11. Follow the steps below to gain a solid working knowledge of AC sweep analysis. When the schematic capture is complete, it should appear similar to Fig. 6.17b. Notice that because ground is defined, there is no need to bring the lower wire out from the schematic to define V_{out} between two points. It is now the single point to the right of the opamp.

6.10.1 *Open a Blank Schematic*
- Open Ltspice; from the menu, File → New Schematic.
- Save it immediately using the menu, File → Save As. Give the schematic a simple name with no spaces under "Name" (e.g., spice_ac). Save it in your personal data directory for the course.

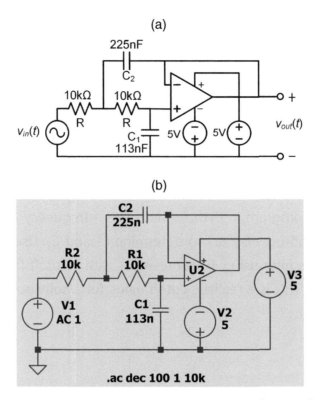

Fig. 6.17 Second-order Butterworth low-pass filter: (a) circuit schematic; (b) LTspice schematic capture.

6.10.2 Add Opamp Integrated Circuit

- Integrated circuits are separated by folders. Choose the [OpAmps] folder and select the "UniversalOpAmp2" component. Typing the first few letters of the name will autoscroll to the right location. If you were physically prototyping the circuit, you could achieve a more accurate simulation by selecting an opamp part number rather than the generic UniversalOpAmp2 part.

6.10.3 Add Voltage Sources

- Click the component icon ⋼. Find the "voltage" component — typing the first few letters will scroll to it — double-click to select it, and place it in your schematic by clicking once. You may want to rotate or flip its direction using Ctrl+R and Ctrl+E, respectively. Right-click the symbol, not the text, to set the opamp power sources to 5 V. You may need to zoom out with the ⊖ tool to place the sources. To indicate that the input on the left is the AC sinusoidal source whose frequency will be swept during the analysis stage, leave the DC value blank, choose "Advanced", and then in the "Small signal AC analysis" panel, enter "1" for AC Amplitude.

6.10.4 Add Resistors, Capacitors, and Ground

- Click the resistor icon ⦚, place two resistors, right-click the component symbol, not the text, and give them values of 10k. Spice understands "k", and there is no need to add the Ω symbol. Do the same for the capacitors using the symbol ⊥. You may want to rotate one using Ctrl+R before placing. The ground is placed with the ⏚ symbol.

6.10.5 Wire it Up

- Use the ✎ tool to wire the circuit together.

6.10.6 Run the AC Sweep Analysis

- Specify the type of analysis to run. From the menu, choose Simulate → Edit Simulation Cmd and choose the "AC Analysis" tab. AC Analysis, short for AC Sweep Analysis, determines how much of the input signal (designated as the AC source) makes it through to the output as the input signal frequency is swept from low to high frequencies.
- To analyze the circuit over a range of frequencies from 1 Hz to 10 kHz, choose the "Decade" type of sweep, 100 points per decade, and 10 and 10k for the start and stop frequencies, respectively. "Decade" means a power of ten, which is most appropriate here, since the desired frequency sweep limits, 1 Hz and 10 kHz, are both powers of 10. This is the standard Bode-style range of logarithmic frequencies studied earlier in the Tech Tip *Bode Plots* following Section 3.2.9 using 100 increasing frequencies per power of 10 (e.g., 100 frequencies between 1 and 10 Hz, and 100 between 1 and 10 kHz). Using more points per decade produces greater accuracy, while using fewer points produces greater speed.
- Choose "OK" and place the text ".ac dec 100 1 10k" anywhere on a blank area of the schematic.
- Select Simulate → Run from the window or choose the ✗ button. The split schematic and analysis window shown in Fig. 6.18 will appear.
- Position the cursor over a schematic wire and click to examine a Bode-style plot of voltage versus frequency. The example shown in Fig. 6.18 highlights the cursor hovering over the wire that is exiting out of the opamp. The cursor changes to resemble a red probe.

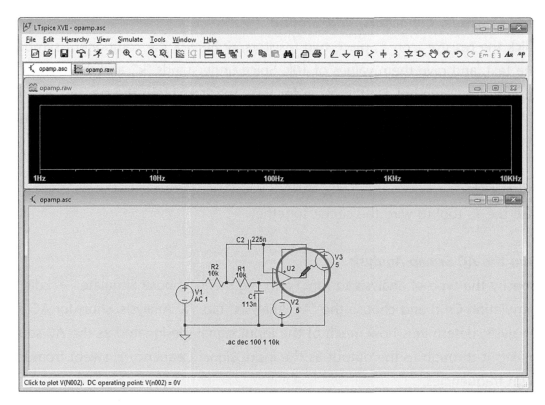

Fig. 6.18 Split schematic and AC analysis window. Placing the cursor over a measurement point changes it to a red probe.

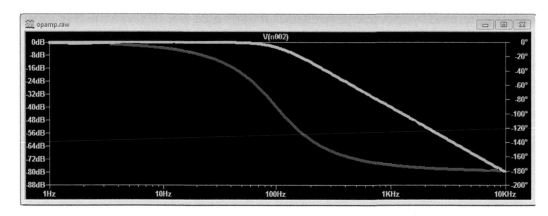

Fig. 6.19 Bode plot showing magnitude and phase. The defaults show magnitude as a solid line and phase as a dotted line; for easier viewing here the magnitude is shown in *green* and the phase in *red*. The magnitude is read using the left axis in dB units, and the phase uses the right axis in degrees.

- Click on the probe to display the corresponding voltage Bode plot shown in Fig. 6.19.
- Two plots are automatically created. One plot shows the magnitude of the signal using the axis on the left, while the other plot shows the phase using the axis on the right. These are known as Bode plots. To remove the phase trace, click on the

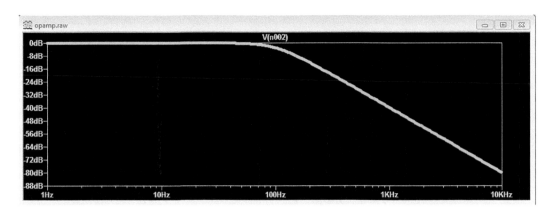

Fig. 6.20 Magnitude versus frequency plot. To remove the phase trace shown in Fig. 6.19, click on the right axis and choose "Don't plot phase".

right axis and choose "Don't plot phase". The resulting plot is shown in Fig. 6.20. Notice how the data shows that low-frequency signals under about 100 Hz are not attenuated. Earlier chapters described how dB units are calculated using a logarithmic scale of $10^{dB/20}$. For example, 0 dB $= 10^{0/20} = 10^0 = 1 = 100\%$ passed at low frequencies. Higher-frequency signals are reduced. The approximately -40 dB of attenuation at 1 kHz corresponds to the circuit passing just $10^{-40/20} = 10^{-2} = 1\%$ of the input signal.

PRACTICE PROBLEMS

6.12 Modify the above walkthrough to use 20 kΩ resistors, but the same value of capacitors. Analyze the performance of the filter over the same 1 Hz to 10 kHz range. Record a screen grab of the analysis window using the Snip & Sketch Tool using Windows Win+Shift ∣ S or Command+Shift+4 on a MacOS. Does the cutoff frequency increase or decrease as resistor values are increased? ®

6.13 The frequency response of a second-order Butterworth filter using the configuration shown in Fig. 6.17a is given by the relation

$$H(f) = \frac{1}{1 + 2RC_1(j2\pi f) + R^2 C_1 C_2 (j2\pi f)^2} \tag{6.6}$$

Write a Python program to evaluate the Bode plot using the resistor and capacitor values shown in Fig. 6.17b, as you did earlier in the Tech Tip *Bode Plots* following Section 3.2.9.

    a. Evaluate the Bode plot over the same range of frequencies as shown in Fig. 6.20, and insert the Python Bode plot into your report. ®

    b. Compare your Python Bode plot with the LTSpice Bode plot shown in Fig. 6.20. Do they show the same results? ®

6.11 ADVANCED TIP: USING NETS

The opamp schematic of Fig. 6.21a is functional but looks messy because of the 5 V sources providing power to the opamps. If we defined a symbol such as $\frac{+5V}{}$ to be 5 V and another, $\frac{-5V}{}$, to be −5 V, then we can significantly neaten the schematic while making the intent clearer. These symbols are examples of *nets* and are illustrated in Fig. 6.21b.

The circuit shown on the right, Fig. 6.21b, is functionally the same as the circuit shown in Fig. 6.21a. The two voltage sources that are apparently unconnected define two nets called +5 V and −5 V, which are used to power the opamp, as shown in Fig. 6.21b. It also uses multiple ground net symbols to simply the schematic further. Practicing engineers much prefer Fig. 6.21b to Fig. 6.21a. The reasons for this may not be evident in this relatively simple circuit, but more complex schematics may have hundreds of powered components; they would be impossible to read if each component had wires connecting back to a single voltage source and a single ground.

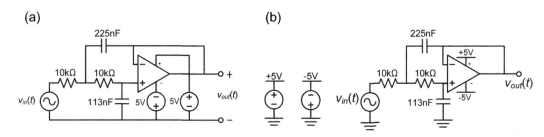

Fig. 6.21 Using nets in schematics. (a) The circuit without nets requires voltage sources wired to the opamp. (b) Using nets simplifies the opamp wiring.

LTspice permits the naming of nets using the ⊕ tool. As an example, to simplify the schematic shown in Fig. 6.17a, first create the schematic capture outlined in Fig. 6.22.

Then click the ⊕ tool and create the net name +5 V, of Port Type "none", and place as shown in Fig. 6.24. Similarly, create a −5 V net name and place also as shown in Fig. 6.23.

Wire them up as shown in Fig. 6.24. Now every time a +5 V net symbol is used, it is functionally identical to being wired to a +5 V voltage source whose more negative end is wired to ground. This is perhaps only marginally neater in this single-opamp example but becomes far neater in more complex circuits that use multiple opamps.

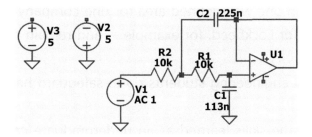

Fig. 6.22 Advanced nets initial schematic. Create the circuit schematic, which includes two power supplies, V2 and V3. Notice the polarity of each supply.

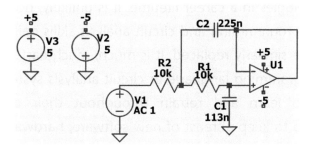

Fig. 6.23 Creating named power nets to simplify the schematic. The two nets, +5 and −5, represent voltage sources that can be used elsewhere in the schematic without requiring long explicitly-drawn wires.

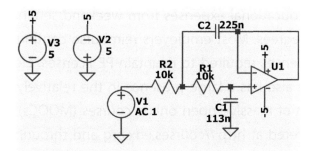

Fig. 6.24 Wiring named power nets to simplify the schematic. Wire the ports as shown to provide the +5 and −5 voltages to the opamp U1 without requiring long, crossed wires from the +5 V and −5 V sources.

In 1962, the economist Fritz Machlup coined the phrase "Half-life of knowledge" to quantify how quickly knowledge becomes obsolete. The outlook is not sanguine for electrical engineers. About a century ago, it was estimated to be 35 years, while a half-century ago, this dropped to a decade, and in 2002, the president of the National Academy of Engineering stated that it was between 2.5 and 7 years, depending on the subfield. Entire career fields are similarly mobile; a generation ago, an engineer could expect to carve out a niche in one well-defined area for one company — missile guidance systems for Lockheed, for example — and remain there for a career. No longer.

What can electrical engineering students do to safeguard hard-won technology skills?

1. Accept that the skills learned as an undergraduate can be generalized and are designed to be transferable. Python and Spice have existed since 1991 and 1973, respectively, but both may be superseded by new technologies in a career lifetime. It is unlikely, however, that the underlying programming and circuit analysis skills gained by learning them will be similarly replaced. It is much, much easier to learn a second programming language or circuit analysis system after the first.

2. Continue to learn and retrain throughout their careers. Use IEEE membership to keep abreast of new software, hardware, and application developments through section meetings and read publications such as the monthly *IEEE Spectrum* that comes with IEEE membership. Many companies offer in-house training programs and reimburse educational expenses from weekend seminars to full university degrees. Most employers reimburse continued education training expenses required to maintain PE licenses. Electrical engineering also has a large presence in the relatively recent introduction of massive open online courses (MOOCs); some excellent ones are offered at http://courses.edx.org and through the IEEE, for instance, with the IEEE Learning Network at https://iln.ieee.org.

Ultimately, it is the same love of learning that drives people to become engineers that keeps them current throughout their careers. Although the profession is changing more rapidly now than ever before, engineering culture is equally quickly adapting to the reality that education is not a college degree but a lifetime process.

COMMAND REVIEW

Schematic Capture Commands

File → New Schematic	Begin a new schematic diagram.
Voltage	Place a voltage source (after selecting, ⊅).
⊅	Place a component.
⟩	Place a resistor.
÷	Place a capacitor.
ʒ	Place an inductor.
ℓ	Place a wire.
↓	Place a ground.

Zoom Tools

⊕ ⊙ ⊖ ⊗	Zoom in, pan, zoom out, and zoom to full circuit.

Schematic Editing Commands

✂	Delete tool.
🖐	Move tool.
🖐	Grab tool.

Analysis Commands

Simulation → Edit Simulation Choose the simulation type:

- Dc op pnt — Voltages unchanging, virtual multimeter (after selecting Simulation → Edit).
- Transient — Voltages are changing, virtual oscilloscope (after selecting Simulation → Edit).
- AC Sweep — Frequencies are changing, virtual Bode plot (after selecting Simulation → Edit).

🏃	Run simulation.

View Results

DC	Hover mouse over schematic for voltage, current, and power. Click wire to label voltage on the schematic.
Transient	Click wire to add to oscilloscope display.
AC sweep	Click wire to add to Bode plot.

Solutions to all starred (*) problems are available on the Elsevier website (refer to page xix for the link).

6.1* Use LTspice to draw the circuit and find the voltage across the 10 Ω resistor in the circuit shown in Fig. 6.25. Hint: Do not forget to assign a ground node in LTspice!

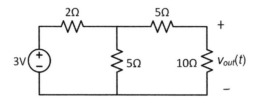

Fig. 6.25 DC voltage circuit. Using LTspice, find the voltage across the 10 Ω resistor (Lab Problem 6.1).

6.2 The circuit shown in Fig. 6.25 is powered by 2 AA batteries. How long will the circuit last if a single AA battery holds 10 kJ of energy?

6.3 Use LTspice to build the circuit shown in Fig. 6.26 and find the voltage across the current source. The current source in LTspice is called "Current" and is found in the same library that holds the "Voltage" source. Print the LTspice schematic showing the voltage at the top of the current source.

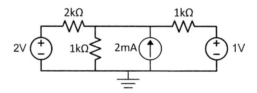

Fig. 6.26 DC voltage and current network. Using LTspice, find the voltage across the current source (Lab Problem 6.3).

6.4* Use LTspice to build the three-dimensional cube of 1 kΩ resistors shown in Fig. 6.27 and find the current through the voltage source. Print the LTspice schematic showing the current through the voltage source. Some versions of the LTspice show current in a wire after simulation by hovering over it. Other versions show only the currents in the postsimulation screen. Take a screenshot of the current through the voltage source. Do not be concerned if the current is positive or negative. The sign indicates only the direction of flow.

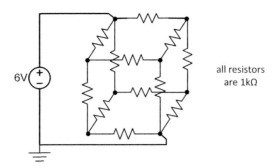

Fig. 6.27 Three-dimensional resistor cube. Use LTspice to find the current through the voltage source (Lab Problem 6.4).

6.5 The *duty cycle* of a square wave is the ratio of the time the cycle is spent "on" divided by the total length of time of each period. For instance, if the on time is 1 second, and the off time is 9 seconds, then the total period is 10 seconds, the duty cycle is 10%, and the frequency is 1/period = 0.1 Hz.

    a. Build two functions in Python. The first, called solve_555, should take values of t_high, t_low, and C and return resistor values R1 and R2 as described in the Tech Tip *555 Timer* preceding Section 6.9. The second, called solve_duty, should take values of frequency and duty_cycle and use them to return the values of t_high and t_low.

    b. Use the two functions you created in part a to determine the resistors needed to build an astable multivibrator 555 circuit with a 1 kHz frequency and a 25% duty cycle that uses a 1 μF capacitor. Hint: They should be different, but between 100 Ω and 1 kΩ.

    c. Build and analyze it using LTspice and print out 4 ms of the waveform showing the current through the LED.

6.6 Repeat all three parts of Problem 6.5 for a circuit that oscillates at 100 Hz at a 10% duty cycle. Use a 1 μF capacitor. Print the LTspice schematic and a plot of the first 50 ms of current through the LED.

6.7* Design a second-order Butterworth low-pass filter like that in Practice Problem 6.11 with a cutoff frequency of 60 Hz using 1 kΩ resistors.

    a. Write a Python function, called solve_butter(), to design a low-pass filter. It should take values of the critical frequency f_c and resistor value R and return capacitor values for C1 and C2 as described in the Tech Tip *Operational Amplifiers* preceding Section 6.10.

b. Use the function you created in part a to determine the capacitors needed to build a second-order low-pass Butterworth filter with a cutoff frequency of 60 Hz that uses 1 kΩ resistors. Hint: The capacitors' values should be different, but both between 10^{-7} and 10^{-6} F.

c. Build it in LTspice. Analyze it using an AC sweep analysis from 10 Hz to 1 kHz. Print both the LTspice schematic and the plot of just the output magnitude, not phase, using the Snip & Sketch Tool on Windows Win+Shift+S or Command+Shift+4 on a MacOS. Use the net method discussed in Section 6.11 to avoid crossed wires. Remember that a capacitor value of, for instance, 2×10^{-6} F is the same as 2 µF, and can easily be entered in LTspice as 2u. Hint: If LTspice returns an "unknown subcircuit" error, check that "UniversalOpamp2" was used and not "opamp2".

6.8* Continue Lab Problem 6.7 by not using LTspice but instead using Python to create a Bode plot of its frequency response using a frequency range from 10 Hz to 1 kHz in the same manner as described in the Tech Tip *Low-Pass Filters* just preceding Section 3.3. Plot the magnitude of your Bode plot, label all axes, provide a title, and insert a 600 dpi high-resolution image of the result into your report.

6.9 Implement the circuit shown in Fig. 6.28 in LTspice. Calculate the Bode plot for the high- pass filter for input frequencies that vary from 1 Hz to 10 kHz as described in Section 6.10, *AC Sweep Simulation*. Print both the LTspice schematic and the plot of just the output magnitude, not phase, using the Snip & Sketch Tool on Windows (Win+Shift+S) or Command+Shift+4 on MacOS. Use the net method discussed in Section 6.11 to avoid crossed wires. Hint: If LTspice returns an "unknown subcircuit" error, check that "UniversalOpamp2" was used and not "opamp2". Do not forget to power the opamp as described in Section 6.10 using +5 and −5 V voltage sources.

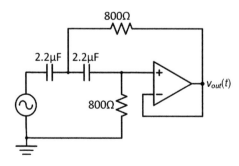

Fig. 6.28 High-pass opamp filter. Use LTspice to generate the Bode plot from 1 Hz to 10 kHz (Lab Problem 6.9).

6.10 Duplicate the results you achieved in LTspice in Lab Problem 6.9 entirely in Python. To do this, you need to know that the schematic shown in Fig. 6.28 is a second-order high-pass Butterworth filter. It can be implemented using the circuit shown in Fig. 6.29, named the Sallen–Key circuit. The frequency response, $H(f)$, is given by

$$H(f) = \frac{(j2\pi f)^2}{(j2\pi f)^2 + (j2\pi f)\left(\dfrac{1}{R_2 C_1} + \dfrac{1}{R_2 C_2}\right) + \dfrac{1}{R_1 R_2 C_1 C_2}} \tag{6.7}$$

Using Python, create a Bode plot of Eq. 6.7 for the component values of Fig. 6.28. To do this, follow the example described in the Tech Tip *Low-Pass Filters* just preceding Section 3.3, but substitute Eq. 6.7 for Eq. 3.3. As in the Tech Tip, define a frequency vector that spans from 1 Hz to 10 kHz using `logspace()`, find the absolute value of $H(f)$, convert it to decibels, and plot it with gridlines using the Bode conventions, including a logarithmic horizontal frequency axis. Do your results agree with those calculated by LTspice in Lab Problem 6.9?

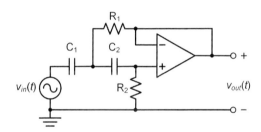

Fig. 6.29 Second-order high-pass Butterworth filter circuit using Sallen–Key configuration (Lab Problem 6.10).

6.11 You are an electrical engineer working for a company that designs natural-gas-powered electrical power plants. One of the newly hired members on your team is upset that the company is not spending more time examining the profitability of generators powered by renewable energy sources, such as wind and solar. The issue was already examined in a lengthy and expensive study a decade earlier, and it was found then that renewable energy sources would not be as profitable for the company in the foreseeable future. The employee argues that the study's goal was fundamentally flawed: that the decision should not rest solely on profitability, but that ethical responsibility to reduce carbon emission should also play a role. Use the five-step method of David Jonassen, with the NSPE Code of Ethics, to analyze the above situation from the viewpoint of the new hire and from the viewpoint of the company's board of directors. Which viewpoint do you find more compelling?

6.12 Use the five-step method of David Jonassen with the NSPE Code of Ethics to analyze the following, unfortunately realistic, situation an engineer may encounter. You are an electrical engineer working for a defense contractor. Your team is completing work on the guidance control system for an antiaircraft missile defense system. Early testing showed that, under certain situations, the guidance control could malfunction, resulting in unstable missile flight over friendly airspace. The guidance control system was modified to maintain controlled flight in that situation. Although the system now passes all tests, you believe that earlier failures may point to a deeper problem with the guidance systems, and that the original flight test conditions could still lead to loss of missile control. However, you cannot be sure of this. All you know is that the system was redesigned to successfully pass the limited number of flight scenarios proposed at the project outset. When you present your concerns to the company management, they assure you that all tests show that the missile system is safe and that further testing not only would be unnecessary but also would lead to extremely costly penalties for failing to produce the device on the contract deadline and might jeopardize the division's, and therefore your job's, existence. They hint that if you fail to approve the system, you may be replaced with a co-worker who would be more willing to approve the project.

You feel torn between your duties to your co-workers, your employer, your family, and the military personnel who will be relying on this technology. You do not know if your concerns are valid without months of additional simulations. What do you do?

INDEX